The Development of Women and Young Professionals in STEM Careers

Tips and Tricks

The Development of Women and Young Professionals in STEM Careers

Tips and Tricks

Michele Kruger and Hannelie Nel

CRC Press
Taylor & Francis Group
Boca Raton London New York

CRC Press is an imprint of the
Taylor & Francis Group, an **informa** business

CRC Press
Taylor & Francis Group
6000 Broken Sound Parkway NW, Suite 300
Boca Raton, FL 33487-2742

© 2020 by Taylor & Francis Group, LLC
CRC Press is an imprint of Taylor & Francis Group, an Informa business

No claim to original U.S. Government works

Printed on acid-free paper

International Standard Book Number: 978-0-367-33440-6 (Hardback)

Library of Congress Cataloging-in-Publication Data

Names: Kruger, Michele, author. I Nel, Hannelie, author.
Title: The development of women and young professionals in
STEM careers : tips and tricks / Michele Kruger, Hannelie Nel.
Description: Boca Raton : CRC Press, [2020] I Includes index. I
Summary: "This book is fluent and systematic. It works through the fears and
ambitions of new engineers in the professional environment. This book encourages
the young professionals and women in STEM to know that they are not alone and provides
insight into their ability to dealing with the stress of developing into a successful engineer
and consultant. This book is ideal for those new to the engineering and consulting fields,
students in engineering education, administrators, libraries, industrial engineering, those
involved in leadership, organization behavior, human resources,
STEM and other areas as well"-- Provided by publisher.
Identifiers: LCCN 2019023872 (print) I LCCN 2019023873 (ebook)
I ISBN 9780367334406 (hardback) I ISBN 9780429322976 (ebook)
Subjects: LCSH: Women in technology I Technology--Vocational guidance.
Classification: LCC T36.K78 2020 (print) I LCC T36 (ebook) I DDC 602.3--dc23
LC record available at https://lccn.loc.gov/2019023872
LC ebook record available at https://lccn.loc.gov/2019023873

Visit the Taylor & Francis Web site at
http://www.taylorandfrancis.com

and the CRC Press Web site at
http://www.crcpress.com

Contents

Preface

The writing of this book came as the biggest surprise to me. I would never have imagined myself to be an author, and life has a funny old way of forcing you outside your comfort zone and delivering surprises. Throughout my career, I have had the honor of being asked to speak/motivate scholars, students, young professionals, and women in the STEM field. Now that I officially train in the field of Women and Youth Development, it has given me a chance to put this book together for you. It's basically all of my speeches put together in a very short "don't make my mistakes, here is what I suggest" guide. I hope that it will change your life and fill it with great enjoyment, fulfillment, and humor.

In it, you will find many tips and tricks to navigate a career in the STEM field, for example, not fitting in is a great thing, and not caring what others think will free your mind.

As I am a big reader myself, it was important to me that this book be short, succinct, and hopefully include a few laughs in between. Enjoy!

I could not have written this book without the help and support of so many people. First and foremost is Dr. Hannelie Nel, my co-author, who not only inspires me every day, but whom I have the honor of calling my friend. Thanks to Dr. Gina Pocock for her input on the book, inspiration, and friendship. Thanks also go to the team from Effecting Change; every one of you has taught me so much, and thanks for forging the path for this book. Thanks go to my colleagues at CSVwater. You taught me that any challenge can be overcome, especially with a dash of humor. Thanks to Mokgosi and the Nextec family for inspiring me to make today just a bit better than yesterday. Thanks to all the support and opportunities from the FIDIC family; I am so honored by all that you have done for me. Thanks to all the inspiring businesswomen in my life, my fantastic and supportive friends and family (including my biggest fan, my sister Charmaine, and the best in-laws anyone can have), and of course my wonderful husband Gerrie and feisty girls Chloe and Isabella. You make my life beautiful!

Michele Kruger

Authors

Dr. Michele Kruger completed her bachelor's, master's, and doctorate degrees in civil engineering at the University of Johannesburg (UJ) in South Africa. Both her master's and doctorate degrees focused on aspects of municipal water treatment. She is a professionally registered engineer with the Engineering Council of South Africa, where she also previously served as deputy chair of the Communication and Information Committee.

Michele specializes in water and wastewater treatment. She is currently an associate director at CSVwater where she is in charge of various water and wastewater treatment projects. CSVwater forms part of the Nextec group of companies. The company profile can be viewed at www. nextec.co.za.

Michele is a former board member of Consulting Engineers South Africa (CESA) and has been involved on various CESA committees including the Marketing Committee and chairing the Quality and Risk Committee. She has been part of the FIDIC (International Federation of Consulting Engineers) committees since 2007. She chaired the FIDIC Young Professionals Forum, served on the FIDIC Sustainability Committee and chaired the FIDIC Capacity Building Committee. She is currently a mentor for the FIDIC Young Management Training Program that develops the top young engineers from around the world, and as such, she is an FIDIC accredited trainer in business development. She helped launch the CESA Women's Forum and chairs the FIDIC Diversity and Inclusion Task Force.

As a young engineer, Michele won the Best Paper by a Young South African in the field of water science and technology, awarded by the UK Federation for Water Research, for her doctoral research. She also won the title of "Young Engineer of the Year" and was later recognized by the Women in Civil Engineering Award sponsored by *Building Women Magazine*, ABSA, and the Department of Public Works of South Africa.

In 2017, Michele was invited and appointed as director of Effecting Change (Pty) Ltd. Effecting Change was established in 2017, and the team comprises successful entrepreneurs, dynamic and visionary leaders, and industry experts in a multitude of relevant fields. Sharing a common purpose, the team strives to facilitate the development of future leaders and entrepreneurs, with a special focus on women and youth. The company profile can be viewed at www. effecting-change.com.

Michele is also conducting research in the field of women in engineering for the Postgraduate School of Engineering Management at the University of Johannesburg (UJ).

Michele has a great passion for engineering and the development of young talent and women in STEM and hopes to make a difference through giving back.

Dr. Hannelie Nel is managing director of Tennelli Industries (Pty) Ltd., a company that focuses on innovation in engineering, technology, and education. Tennelli addresses critical concerns inherent in current design and optimization practices, while supporting people-centric issues impacting industry and finance by focusing on value creation in the areas of cost and quality systems. Hannelie is a registered professional engineer with the Engineering Council of South Africa, and her company profile can be viewed at www.tennelli.com.

In May 2018, she was nominated and elected as a non-executive board member of Denel, a state-owned company of South Africa under the auspices of the Department of Public Enterprises and Minister Pravin Gordhan. And in July 2018, the Southern African Institute for Industrial Engineering (SAIIE) awarded her with lifelong honorary fellowship in recognition of her contribution to industrial engineering.

In 2017, Hannelie was invited and appointed as director of Effecting Change (Pty) Ltd. She also serves as senior academic and researcher in the Postgraduate School of Engineering Management at the University of Johannesburg (UJ). She has published over 40 academic peer-reviewed papers, including books, and has delivered several keynote addresses. She holds a DEng in engineering management, an MSc in industrial engineering, and a BEng in chemical engineering. She has 20 years of experience in business and academia and was a former vice dean of the Faculty of Engineering and the Built Environment at UJ. Hannelie is the founder and current ex-officio director of the UJ-Group Five Women in Engineering and the Built Environment (WiEBE) Programme. WiEBE was established to support, empower, and advance women in engineering and the built environment; and to raise awareness of gender parity in companies. The non-profit organization can be viewed at www.wiebe.co.za.

In both 2017 and 2018, Hannelie was nominated for the prestigious Professor Kris Adendorff Award for Prominent Industrial Engineering Professional by the Southern African Institute for Industrial Engineering (SAIIE). She is the former president and an honorary fellow of the SAIIE, and serves on the board of the South African Society for Engineering Education (SASEE) and the TechnoLab at UJ. Previously she served as strategy manager and board member of the Metal Casting Technology Station for eight years, a technical Centre of Excellence established by the Department of Trade and Industry through the Technology Innovation Agency (TIA) for the foundry industry in South Africa. During that time Hannelie applied for and received funding to establish a second technology station in Environmental and Process Engineering. She also established an international dual master's program with Freiberg University in Germany.

In 2017, Hannelie was invited to join the International Women's Forum, a global organization in support of senior businesswomen. She joined the Institute of Directors in 2016 and is an international author and speaker with two published books to date. Notably, her book titled *Leadership and Agency by Women Engineers in South Africa* was published in 2015.

In 2014, she was selected as finalist in the Standard Bank Top Business Women Awards. Other awards include the 2013 WiEBE Award for Outstanding Contribution to Women in Engineering; the 2011 IEEE WIE Award for Human Capital Development in Science, Engineering, and Technology; and two SAIIE awards for the promotion of industry and academia liaison.

Passion for the Job

1

1.1 FORTY YEARS (OR MORE)

One of the most important things to realize is that you only have one life on earth. One. Let me say it again... ONE. This is it; there is no practice run, no rehearsal, just this. No matter what religion you practice, you still just have one life on this earth. Ok, I stand corrected ... in this specific form, if you believe in reincarnation.

What does that mean? It means that you need to rid yourself of the notion that having that dream job/life is for "others": the famous, perfect "other people" beautifully portrayed in advertising enjoying amazing ski trips and sun-kissed beach holidays, with beautifully perfect children—going from success to success. I just need to point out here that those people are models/actors (who probably don't even have children), interacting with children they just met that day. IT'S NOT REAL!

Now, with that tirade done, let's get back to you. There will ever only be one awesome you. Only you will understand yourself best and what is best for you. Not your parents, not your partner, not your boss or friends. So now that we have established that you only have one life and only you know your strengths and weaknesses, what you like or dislike, it is now up to you to imagine your Dream Job.

The Dream Job doesn't have to remain a dream. It can be real if you systematically move yourself in that direction. That said, your idea of a Dream Job will change as you go throughout life. As it should! We all change and develop as the years pass.

As you start your career, often at only 18 years old, you are usually in a situation where you have no idea what you want, but you have to make a choice, NOW! You then spend an average of three to four years on study and training, which dictate the rest of your life. I just want to point out something profound—3 years are dictating the following 40 years or more ... and your happiness. In an era where we are now told the first person who will become 200 years old has already been born, this is actually quite ridiculous. After studying, we are traditionally conditioned to think that the next step is to find a job, slog through, hope for promotions and raises, have a family and through further slogging, create opportunities for your children. Through even further slogging, you assist in raising grandchildren, and if you are really lucky you get to enjoy that thing you have actually been waiting for your whole life—retirement. Once again, glamorized by the media, retirement is shown as two beautifully graying, healthy, fit people taking long walks on the beach and lifting grandchildren in the air. The truth is often very different as retirees are struggling to pay escalating medical bills with savings far less than that promised by another media campaign put together by pension companies.

Instead of feeling alarmed by this, I want you to start feeling excited! "But Michele!" I hear you say. "You just freaked me out and depressed me! I will get extremely old and I won't have enough money to do so!" What I am saying is: if you get your head right, you won't want to retire and you will have no problem financing your old age. Not only that, but you will love every minute!

Instead of slogging through the next 40 years or more, I want you to imagine yourself as if you were a child running into a new playroom. There are so many toys; you can't wait to play with them all! You choose your favorite first and then move on to the next, then the next. There aren't enough hours in the day to get to all the toys!

As in life, your playing is disrupted—you need to eat, have a bath, you need to do homework, you need to play in a group game. There will be bumps, but overall it is a wonderfully enjoyable experience. In life, I equate the bumpy road to family celebrations and tragedies, family fights, having children, and changes in economic circumstances. But overall, life should be very enjoyable, not a slog. A career is not a 40-yearlong slog at one company, being grateful for a salary and a life unfulfilled. Your life and career are an endless and unlimited buffet of experiences, to be sampled and enjoyed (or spat out sometimes). And the only person who can create this for you is you. Nobody owes you anything. Nobody cares more about your life and career than you. Shouldn't **you** then be in the driver's seat deciding the road and destination? Cue roadtrippin' music here…

1.2 "KNOW THYSELF"

One of the most jarring moments of my life was a scene that played out in the movie *The Matrix* where the Oracle points out to Neo the meaning of "Temet nosce"—"Know thyself."

The phrase was previously touched on by the philosopher Socrates who taught that:

"The unexamined life is not worth living."

Whoah! Mind blown.

Do you really know yourself? Have you examined your life? The good, bad, and ugly? Are you able to be really honest with yourself in all aspects of your life?

When I heard these words for the first time, I was relatively young and it was hard to admit to myself that I obviously had weaknesses, as I was fighting to make something of myself every day. I was fulfilling my destiny, wasn't I?

Looking back, I was definitely doing the right thing for me at that time. I just never saw myself at that time moving to where I am now in life. I was in such a bubble, indoctrinated with the needs of everyone around me. As you move through life, you will constantly need to examine yourself and your life and adjust your buffet sample plate accordingly. Are you still on your road, or are you taking a detour on someone else's road (in other words, doing what they need you to do)?

I challenge you to go sit somewhere quiet, preferably with an amazing view, and be honest with yourself. Who are you? The real you? Look at your life from all angles. Is it going where **you** want it to go? Not where your parents, partner, boss, or where you think society wants you to go? Are you being the real you?

Ok, now that you have that down, I don't want you to change yourself. I want you to **be** yourself. I want you to love that person who is there deep down, warts and all. Not the front/pretense you put up as boss, worker, parent, child, friend, or community member—the real you. The bigger the gap between the real you and the person you let others think you are, the more energy you are spending on just existing. No wonder you feel tired all the time!

Just be yourself! Spending a lifetime pretending to be something or someone else is pointless and will eventually make you bitter when you are not rewarded for this unconscious effort. Silly you!

The more time you spend being yourself, the more you will attract like-minded people and the more time you will be spending with people you really enjoy! Do you now see how you have actually been sabotaging your happiness by pretending to be someone else and surrounding yourself with people who don't fit with you? Or even worse, irritate or bully you?

Exactly. Know thyself. Boom!

1.3 ARE YOU A PENGUIN?

"What?" I hear you say. "Why would I want to be a penguin?" Well, it's not so much being a penguin as being judged as a bird when you are a penguin. Imagine this: You have all the tools a bird has—you have wings, you lay eggs, you have a beak—but alas you can't fly, you look ridiculous walking on solid ground, and chasing after food is impossible. You feel absolutely useless, and every day you are being judged at how bad a bird you are.

Now, put your penguin butt in water, everything changes! You fly through the water and you are elegance personified! You are at home in the water and finding food is ridiculously easy as you shoot like a speeding bullet through the water. Not so useless now, are you?

A career in STEM (Science, Technology, Engineering, and Math) can be the same. There are historic expectations placed on you, that the perfect STEM person should be so and so and so. You may be placed in a box with certain expectations, and even though you are killing yourself to make others happy, you just don't fit and sometimes you just feel useless (and often afraid).

Well, no more my beautiful penguins! You can bring your passion to your STEM career and fly through the water. Diversity is the key to success! Having only one personality type, just a bunch of mini-mes of the boss is a sure road to business failure (how can there be new ideas if everyone is the same!). Having diversity and multiple personality types living out their passion means that you will all bring synergy to your environment and to the company. It takes great leadership to recognize when someone is feeling like a penguin on land and how to get them into the water, but more on that later. However, it's up to **you** to go to management to tell them about your penguin problem and it's **you** that has to convince them of your passion and talents and how getting your butt in water will take the company further (and be more profitable/productive).

An example of this is when you are required to do highly technical, repetitive work (such as laboratory testing). You are bored out of your mind and you need to have contact with other people. Is there then perhaps an opportunity for you to interact with clients? Will your warm and charming personality not be better suited to that rather than that of the highly introverted manager who prefers not to deal with people? The poor manager loved his job in the lab and now a promotion means he is in client liaison hell.

The opposite can also apply. The worst thing you can do to a highly productive and profitable engineer/scientist that prefers detail design or research and development is forcing the person to do cold-call business development. These people hate making calls, so this is their personal hell!

So speak up, penguin! You have so much more to give! And you don't have to wallow in hell, just convince your manager to shove you in the water.

1.4 PERSONALITY TESTING

My biggest tip in the world to you is to do personality testing. Some are free on the Internet, and some are hugely expensive. It doesn't matter; do as many as you can! You, my dear penguin, will surprise yourself! And find yourself.

1.5 FINDING YOUR PASSION IN YOUR CURRENT SITUATION

Now that your career can be viewed as a wonderful buffet, I challenge you to consider how your current job can be viewed. Is it a tasty morsel, more than the sum of its parts in an exciting taste combination shown so delightfully in the Disney animation *Ratatouille*? Or is it a dry piece of steak—fulfilling its duty to nourish but could do with something extra, like spice or sauce? Or worse— boiled, bitter Brussels sprouts?

Life is what **you** are making of it. You may be stuck with a steak right now, not of your making (you would have made it nice and juicy). How can you use your resources to add a nice sauce or spice it up? Why not add a lemon butter sauce to the Brussels sprouts?

Let me use an example. You are currently working for a big corporation. Your job is ok, you are able to create security to your extended family, and in general, you have nothing to complain about. Every now and again, you feel pressurized by deadlines or having to find projects to keep the company going. You have staff to manage and there will always be HR issues to deal with. Not a Michelin star dish, but like meatloaf, it's ok on a weeknight.

If you are not able to make drastic changes, what will improve this situation? **Finding your passion**. If you can find purpose in your work, it won't just be a slog; it will be a reason to live. As an engineer myself, delivering water and sanitation solutions to the people and industries that so need it, I am not only contributing to the underprivileged who need the services to survive, but I am also part of the great wheels turning to make the economy tick over.

I hope to use my skills to make local businesses more competitive in the national and global economy.

Accountants and auditors fight the good fight against corruption, lawyers protect individual and corporate rights, and utilities literally keep the lights on for schoolchildren to study by and improve their lives, not to mention the contribution to the economy.

Now that you view that meatloaf a bit differently, you can spice it up. Instead of only "showing up" to work, "doing what is expected," and "fulfilling your key performance indicators" in order to make a salary, how can you leave this world just a bit better today than it was yesterday?

Have you noticed any inefficiencies or possible improvements in processes you are involved in? How would you approach the problem if the department you were working in were your own business? In a roundabout way, it sort of is … but more on that later.

Challenging yourself to improve processes and the customer experience as well as understand the financial side of things are both ways to add spice to your corporate life. Imagine yourself as a customer of your company. Would you pay what customers are currently paying for your goods or services? Would you pay less or more? Why? Will they keep on doing so indefinitely if nothing changes? What is your value add? What sets you apart? How can these challenges bring an additional dimension to your working life? How can these experiences shape the road you want to follow? Shape you?

On a personal note, I suggest you add opportunities to incorporate your hobbies into your work. For my job, I travel extensively, which I love! Even though I am not the best cook, I love watching cooking shows and I appreciate "foodie"-type foods (hence my constant references to food). Whenever I travel, I enjoy local cuisine. The stranger the food, the better! I often remember a place not per project or landmark, but by a great dish I enjoyed there. Is there anything that you enjoy that can be incorporated in your work life? Just try, what do you have to lose?

What if you have tried all of the above and you are still not happy? Still not finding purpose, passion? You have three choices:

- Suck it up and muddle through
- Get creative and find additional ways not mentioned above to make the experience enjoyable
- Leave

Yep, I said it. At some time, you need to man up and leave (or put your big girl/boy panties on). If you have a narcissistic boss, you will be bullied no matter how hard you try or how much money you make for the company. At some point, you need to understand that there will inevitably be problems everywhere you go, but there may come a time in your life when you are ready for new challenges. Time to take the road less traveled.

1.6 FINDING YOUR PASSION ELSEWHERE

Now that you have decided to leave, the big question is where to go? Well, step one is to let the most important word guide the way: **passion**. What is your purpose in life? If you are leaving the current company to join a similar company/job description, it could be because of a certain person (normally your manager) or the company culture. If so, revert to the previous section for advice. If you are brave enough to leave to start your own company or to work in a different field (this could even be a different division at the company), welcome back to the buffet table. Time to try out the next dish! Personally, this chapter of my life is a bit of a hybrid as it hasn't meant me leaving my company or changing my role. I am assisting the development of new products, involving additional business units in the holding company. Talk about taking up an exciting challenge! But back to you.

Your bravery will only pay dividends if you do not revert back to the same old. You have learnt many things at the old company/division: how to do things and how not to do things? What are the value and limitations of red tape? What is the importance of financial literacy? How do you now take your passion and experiences and make a success out of things?

As discussed before, step one is finding your passion or purpose in life. And no, **making a profit is not a purpose**, but a means to an end. If you are passionate about making life just a bit better today than it was yesterday, then profits will follow. What ability do you have to make life a better place? Are you an amazing teacher? Listener? Flower arranger? How can you incorporate your skills with a business? What is your niche? (Check that personality test you did!)

One of the most important questions in our age is to ask—How can you incorporate this with technology? Is there an opportunity for an app, artificial intelligence, virtual reality, big data or business analytics? Keep in mind the demographic of your customer and the fact that most people have access to knowledge via the Internet on their mobile phones. How does this affect your business plan? What is unique to your proposition? Is there already too much competition in this space?

Once again you need to put your hands on the wheel and envision the road. The destination is a bit fuzzy, but you have an idea which direction you are headed in. This too may change later, and you know what? That's ok, too, just as long as you are still the driver.

1.7 FINDING A NICHE

1.7.1 Definition of Niche

A niche is defined as a specialized segment of the market for a particular kind of product or service. Per definition, it implies that not many others can provide this service. If you are able to provide a niche service at your company or in the market, you have very limited competition, creating a larger opportunity for yourself or for your business. What is your niche? How does it correlate to your passion?

1.7.2 Creating a Niche

Opportunities are created when a problem is solved. You can have the best idea in the world, but you will not be able to deliver a fit-for-purpose product if you don't fully understand the problem experienced by your future client.

There are problems already covered by many excellent service providers, for example the need for mobile telephones.

What can you or your business provide that is still very much needed but has very limited service provision or needs specialist input?

In business or as an employee in a large corporation, this usually requires following the road less traveled. It means providing a service that needs that extra input/ingenuity/effort, offering to master difficult software, or offering to work with a notoriously difficult client. If it was easy, everyone would do it. Remember that.

1.7.3 Finding Your Niche

What to do—play to your strengths. It may seem obvious, but this goes hand-in-hand with knowing thyself and finding your passion.

What not to do—make your life a misery. If you are walking down a road less traveled, it may create a niche (working with difficult clients), but if you typically avoid conflict or do not know how to handle difficult people, this road will be filled with misery. If you are doing something every day that you hate, you will suck at it. Yeah, I said it! No niche created. You are just being a penguin afraid of water.

Here are some ideas to start with when trying to figure out your niche:

- Are you good at connecting people or networking?
- Are you able to see solutions where others struggle to see past the problem?
- Are you good at seeing how unrelated things fit together?
- Are you good at seeing how things will be improved with the use of technology?

Sometimes you may not recognize your strength as such because it seems "so easy, surely everyone can do that too?" Ask friends, family, and colleagues what they think you are good at. If you hear something specific from more than one person you can investigate further.

1.8 OTHER PASSIONATE PEOPLE

In life you will have people who will give you energy and those who suck it from you. Coming back to our food analogy, there are waiters that will describe the specials for the day so successfully you can almost taste the smoky flavors and underlying berry tones. Then you have waiters who serve you with a screensaver look, casting a shadow on the food created with dedication inside the kitchen.

At the table you have dining companions who will delightfully share their food so that you too can taste how wonderful it is, their faces showing their intense enjoyment of the moment. Then there is always one who concedes that, although the food is great, there is always just the one thing not to their liking. The food is too this, or the service too that.

In life I have met people who, despite having overcome incredible hardship and pain, remain positive and are sampling experiences of life with joy, doing it "their way." I have found that, incredulously, this is not **in spite of**, but **because of** the challenges they have faced.

For a long time, they were not the driver. They were snapping at scraps falling from the buffet table. Thanks to those challenges, they now not only appreciate the buffet table so much more, they too want to leave the world a better place, giving back, hoping that they can assist in others finding the joys of the buffet table.

They live their passions, gratefully so, knowing full well that life is not fair and it all can be taken away at any moment. They know that they are not owed anything by anyone, and that is also what sets them free. Whoah!

So, surround yourself with such people. Combined with your passion, the synergy created will lead to amazing experiences.

1.9 ROLE MODELS

1.9.1 Inside Industry

I have been honored to know amazing role models in my industry. These role models can be viewed from many points of view. I have female engineering and science role models who fight the good fight on a daily basis to turn around gender-based bias and promote gender parity. These women have shown that success is attainable as both entrepreneurs as well as intrapreneurs in large corporations. Although you may feel you are alone, there are other women who are going through the same situation and winning!

From an engineering point of view, I have been honored to know amazing people in my field, pushing the edges of science and technology and embracing the digital world. These people's fearlessness is what inspires me the most. They do not see failure as a bad experience, just a sharpening of their pencil. To them, the real failure is not trying at all, and apathy is a swear word. These people not only see improvement of the status quo as a challenge but see business or personal growth opportunities everywhere.

1.9.2 Outside Industry

I strongly encourage you to find role models outside your industry as well. Understanding these role models will assist you in combining your talents with other industries to create innovative ideas based on synergy. What made these other business leaders a success? Can this be incorporated in your industry? How do these role models handle the stress? How do they create opportunities? What do they do for fun?

Speaking of which, every day it becomes more evident what an amazing person Nelson Mandela was. Besides the obvious achievements known around the world, I don't think we ever comprehended how difficult it must have been to be president of South Africa at that time. He not only had all the fears and expectations of the world on his shoulders but he also had to manage the fears and expectations of the beautiful rainbow nation he had worked so hard to unite. Listening to his international interviews showed me that you can be a strong person and do it all with quiet grace and dignity. What a wonderful person!

1.9.3 Global

There are many international role models constantly referred to. One of my favorites is the controversial Elon Musk. I may be biased, as he was originally from South Africa, but if ever there has been a person who could successfully let go of the cloak of "but this is the way we have always done it," it's Elon. Not only can he make dreams come to life, he has successfully shown he can do it cheaper, faster, and more sustainably.

Ironically, the controversy surrounding him makes him more human and real to me. For a long time, he seemed superhuman in his achievements, forsaking a personal life and family to bring the world what it needs, seven days a week. Talk about living for your passion! But it can take its toll, as seen with Elon.

I am a strong believer in a balanced life. As history has shown, it can all disappear very quickly. Despite having given his all to Apple, Steve Jobs lost the company he built from the ground up, regardless of his efforts. He later enjoyed "revenge" by once again assuming leadership, and Apple has achieved unprecedented success due to his input. In difficult personal times, that's where the "other" parts of your balanced life come in to save your sanity.

Another international role model who has been able to successfully "have it all" is Sheryl Sandberg, COO of Facebook. Her book *Lean In* has made a huge impact on my life, and I encourage every female professional to read it. Her book has also, in part, inspired me to write this one. The fact that she concedes that she had to overcome doubts in writing the book is inspiring, including "but how could she not?" as she stated in her book. She succinctly approached the fears we face every day with practical guidance to boot. Don't wait, read the book ASAP! Surely all she has gone through has helped her to survive the struggles that Facebook has subsequently faced. A side note here— her life shows that there will always be bumps in the road, even after everyone thinks "you made it." Remember this!

1.10 SHUNNING FAMILY/PEER PRESSURE TO MAKE THE RIGHT DECISIONS FOR YOUR CAREER

You are not your friends or family. They are living in their own bubbles and, first and foremost, any decision you make will be perceived through the lens of what effect it will have on them (and their bubble). As a young person, this could be difficult, as any parent wants the best for their children. Parents may feel that they have failed in achieving their own dreams and feel the need to live their dreams through you.

No one doubts that you love your family very much and want to make them proud, but you are the one who will be doing this for the next 40 years or more. If you are not able to break free from your family's pressure, consider the fact that although this first part of the journey is not of your choice, hopefully, you will be sampling as much as possible from the buffet of life and that this first part is just a starter and part of the learning journey. Every experience can be viewed as a learning one, part of the personal knowledge arsenal you are creating for sampling the buffet.

If you are much older, the first response will be around how either the loss of your time or income will affect your family. These are very valid concerns. Losing your house or not being able to afford school fees is not on anyone's wish list. This is where proper planning and being realistic comes in. Do your homework, make sure you have everything in place to reach your goals, and then defend your business plan! But most of all, save as much money as you can before starting a new endeavor.

If you are currently part of a company and wish to change the focus of your work, don't forget: your boss and colleagues will guide you where **they** need you, where the **business** needs you. Ask me, I know how badly this can go: There I was assisting in water issues for the mining division, and the next thing I know I was being told I was no longer a water engineer, I was now a mining engineer! Shouldn't this have been my decision? Push back! This is **your** life; don't compromise your career path. If not, the next thing you know, you are on a path dictated by the company for a few years, then a few decades, and so on.

So let your passion, personality, and talents guide you, having **you** and only you in mind, because it will be **you** delivering on this—for a long, long time.

Ultimately, this is also the best course of action for any company you work for as you will not be as profitable and productive as you can be when following a path not meant for you. To repeat, my friend—you will suck at it—which in turn will kill your spirit.

Emotional Intelligence 2

2.1 SO WHAT EXACTLY IS EMOTIONAL INTELLIGENCE?

In my personal experience (you are welcome to textbooks if you want the technical version), emotional intelligence (EI) is putting yourself in someone else's shoes.

2.2 WHY IS EI IMPORTANT?

2.2.1 Subordinates

If you are able to put yourself in someone else's shoes, you are able to try to feel what they feel, good (yay, I just had a baby) or bad (what is sleep, can you eat it?).

If you are managing staff, the above will obviously become very important. You **have** to celebrate the big moments in the lives of your subordinates! They need to feel you care. They too are living their one life and need to feel that you don't just treat them like a number.

In an industry such as STEM, EI can be your trump card! Let's not kid each other here, traditional STEM personalities can be **very** EI deficient. Sorry penguins, but I have had some serious humdingers!

My worst experience was when I heard a manager say, "I don't have to give credit to my staff ... that is what they get salaries for!" What? No dude ... just no!

2.2.2 Clients

Your clients need to feel that you care! About their deadlines, their KPIs, their status. Your clients need to feel that you can imagine yourself in their very (politically sensitive) position. You need to make them look good and give them credit in front of their managers if they have impressed you. They deserve it! Because in their very politically sensitive positions, praise happens rarely and most often their superiors are not as technically specialized as they are and don't recognize their technical achievements.

2.2.3 What Is My Version of Success?

At a young age we envision ourselves with a certain life, certain material possessions, maybe a family, but most of all an infinite degree of satisfaction and independence (by the way, so does everyone around you, my EI-savvy penguin). The reality is not always exactly what we would have hoped, but at some point we feel to a certain degree—yes, I made it! Oh, and chocolate. Random but there it is.

Well, what does that look like?

- Have I provided a steady income to put food on the table?
- Can I afford schooling for my family?
- Can I even afford holidays, that thing that is so elusive for so many?
- Do I feel satisfied and fulfilled?
- Can I afford chocolate? (Yeah, had to slip that in there.)

2.2.4 What Is the Meaning of Success to Me?

If you read nothing else in this book, please read the following:

Success is nothing you thought it would be. And that's a good thing.

When I was young, success meant money, as we struggled (a lot) financially. Now, I find success is actually keeping it all together, one day at a time, dropping the occasional ball as a mom, but keeping myself and my family alive and (relatively) happy and being able to keep a roof over our heads and our tummies full with the help of my husband. And chocolate of course. Compared to most third world nations, this is excellence! Being able to travel and go on holidays is just the cherry on top. Having a great job, training young people and women, and exercising the other interests I enjoy, gives me self-actualization.

What does success mean to you right now? Whatever it is, make sure it's **yours**, not what others are trying to imprint on you. Why buy that BMW when you don't even like the car? And, be assured that your version of success will change, just like you, my penguin. You may start to like dark chocolate!

2.2.5 Playing to Your Strengths

From a young age, we are conditioned to not blow our own horn. We're taught that arrogance is annoying. The sad truth is that if you are not going to point out your strengths, there is a good chance that no one else will be doing it for you.

But hold on! Before you start annoying others while describing your endless strengths, have a long, deep think about what these strengths may be. Don't be shy, but be realistic. These have more weight when they come from a third party, so ask everybody you know what they think your strengths are so that you can build on them. Offer these as carrots to your employers (in a non-arrogant way) and then prove yourself! Use them, build on them, and set yourself apart!

Later in life, you will use these strengths to find fulfillment, generating money from them or just enjoying them as part of a balanced life. Later in life, you will also learn that it is imperative to know your weaknesses, that you should be honest about them and never promise something you know you cannot deliver. There is no shame in this, it's just good EI.

2.2.6 Being Needed (Which Is What Most People Want)

This is one of the most difficult but also one of the easiest topics to discuss. It is difficult to admit that we need to feel needed, but once we have done so, it actually will bring so much to your life. I want to feel needed; therefore, I need to contribute so that my contribution is invaluable/appreciated. So, what do I do now?

Try to see it from the alternative point of view. Imagine going through life feeling that no one would be affected if you existed or not. Ok, besides the taxman … but somehow, I never really feel appreciated by him. Sorry, back to you. Only feeling missed by the taxman is terrible. So instead of feeling like others should spontaneously start appreciating you, think of the famous American phrase, "What can you do for your country?" A phrase which can easily enough be translated to "my town, job, boss, industry, family, friends, peers, youth, and so on." All of a sudden you can see there is definitely someone out there that needs you, my friend. Most definitely!

2.2.7 Putting EI to Work for You

2.2.7.1 First and foremost ... don't give up your power!

Each and every one of us feels powerless at one time or another—and all too often such feelings arise in the workplace. But you also have choices—even if your choice is to stay and do what's being asked of you, such as being nice to a boss you who drives you crazy or being pleasant to a co-worker who is always rude.

It's no small thing for you to recognize you're making the choice—consciously and intentionally. Not responding is also a powerful response.

You gain nothing from feeling as if you're a victim of your circumstances. This may seem like a small point, but if we realize that in choosing to accept our circumstances, then we are in fact making a choice to do something, and we release some of the powerlessness of feeling trapped.

When you feel trapped, your power shrinks—as does your ability to see solutions right before your eyes. Give yourself credit for making that choice—and for the strength it takes to do just that.

2.2.7.2 Build strong relationships at work

Start to build a strong support network from the first day you begin a job. A good tip is to see everyone as someone you can build a working relationship with. Some people will resist your attempts based on their own agendas or views of the world. Focus on those relationships that you can build by being helpful and supportive.

2.2.7.3 Your attitude at work

If you come in with a sour attitude or always see **them** as wrong and you as right, then you are the one who loses out. Attitude is a matter of perspective, and perspective matters. Try telling the story from someone else's eyes—a powerful way to re-frame what's happening so you can start to see ways you might change your game plan or be able to help.

2.2.7.4 When emotions hold you prisoner

Sometimes, for whatever reason, you find yourself getting caught up in a rush of anger or frustration. At times like these, remember your power to choose and help change things for yourself. Be aware of things you might have done or are doing to set yourself up as a target.

2.2.7.5 Beware the blame game

The blame game is when you point your finger at everyone and everything except yourself. You lose precious time and perspective that could be helping you create a more enjoyable experience for yourself. Both acknowledge and discuss your feelings (preferably with a friend or therapist). People react to who they see and not who you really might be—or what you could offer them. What they see is what (impression) they get. Stop blaming and be known as the fixer instead of the blamer.

2.2.7.6 Being right just doesn't matter

You're focusing on the wrong things and you'll only wind up diverting yourself from getting ahead. No one likes a smarty-pants: actions speak louder than words. Prove your worth not by being right all the time but by being someone who helps get things done and who solves problems.

2.2.7.7 Seeing possibility

When we stop focusing on anger and hurt and all that we think is missing from our work lives, we open up to seeing real opportunity. Possibility is all around us if we just learn how to look for it.

2.3 HOW MUCH DO YOU CARE ABOUT WHAT OTHERS THINK?

I gave this topic an important heading as it's one of the most important lessons in life—ever.

The best advice in life I can ever give you ... **EVER** ... is: do not care what others think of you. Not your family, not your friends, not your colleagues, not strangers. Arguably these people are in their own bubbles, trying to make the best impression they can on those important to them and then projecting on to you their desires for success and achievement. Are you (again) just buying that BMW (you can't actually afford) just to impress your (materialistic) acquaintance? Are you buying a house you don't really like, just because a friend described that they would like something similar? Are you spending money on a holiday in a place you don't really want to be just to get social media "likes?" The only people who you should spend your energy on trying to impress are yourself and the people in your life who would have your back at all times. They don't care what you look like in your pajamas; they can see your heart. They truly love you. Do not ever sacrifice yourself just to get someone's approval/fit in/keep up with the Joneses/or for profit. That, my friend, is just chasing the wind.

2.4 BE MEMORABLE, BE YOURSELF

I can hear you saying, "Michele, are we still on the EI/EQ tangent?" Yes, oh yes, we are. One the worst disservices you can do to yourself is trying to "fit in." Who in life is going to remember you when you keep on trying to fit in, be like others, emulate your peers?

How is your boss, company, or client going to advance your career when you do not stand out from the crowd? For once in your life, do not fit in, fit out!

I am not asking you to be something you are not, I am asking you to play to your strengths. Disregard what others think, and be your unique, fabulous self. If you want me to promote you, I need to remember you!

I also need to remember you for the amazing job you've done, so give it your all, do your best, work hard, and become that person that will be hard to replace/retrench.

2.5 SELF-ACTUALIZATION

Again, I hear you say, "Michele—self-actualization, is this still part of EI?" Oh, yes, my friend. And the best part of it is, in the weirdest way you will not expect.

If you are not familiar with Maslow's hierarchy of needs—you move up from the bottom, and should you have a knock in life, you move down, until you are able to move yourself up again—look it up.

The first three levels deal with basic needs that are typically fulfilled through relationships, and by securing a meaningful income to allow you to meet material needs. Typically, in the modern context having your "ducks in a row" brings you the first three levels. In other words, you have a job, a roof over your head, food, possibly a partner (and kids), and you have one or two friends to hang out with. They are not necessarily easy or a given, but I believe in you enough to know if you are reading this book, you have enough gumption to be comfortably fulfilling these levels. Romantic love, alas, that is life … and another book.

Self-esteem and self-actualization, levels four and five, are powerful tools to enrich your life when you hit the proverbial "what is the meaning of life?" wall. And yes, it will come. They are not only tools to enrich your personal life but are able to help you find the career path for your professional life.

Being a success in your chosen field/community brings you to level four. But level five, that is where I want to take you, and you can do it! I promise!

Remember, if you don't care what others think, it is easier to surpass the previous levels, as you do not need the best house or car or whatever other status symbol is currently required.

If you are not worried about being good enough or being rejected because of your earthly belongings, or lack thereof, you are more open to embrace the people in your life and embrace level four without prejudice. "What are you talking about Michele?" Ok, just don't place your worth on what others think of you. Better?

The easiest path to achieve level five, self-actualization, is thinking how you can contribute to society and make this world just a bit better today than it was yesterday. Here is the humdinger—without expecting any acknowledgement or payment for it. Just do it because it's the right thing to do. "Sho! No way!" I hear you say. Yes way, my friend. It's in you, even if you don't feel it right now.

For me, it has been speaking at conferences for young and female engineers, giving them tips and tricks to navigate this difficult path. It has been writing this book to assist others in finding a meaningful path in life and enjoying their chosen profession, not just muddling through. You can do it, too—give of yourself, it will feed your soul! I promise!

When you have achieved this level, you feel that your "work" is your love and not a hardship at all, something you can do for 200 years and still feel you need to do more of. In other words, once you have reached self-actualization, you can never stop working, because you will love every minute of it! Mind blown!

Thinking Strategically about Your Career 3

Ok, penguins, now that you "know yourself," you have an inkling about your passion, and you can put EI to use, the next step is to help you navigate your career as a young professional/woman in STEM. This is where I teach you tips and tricks to fast track your career and set yourself up for success. This is the part where you don't have to make all the mistakes I did, but make your own! Then write your book to help others! I dare you!

3.1 YOUR ATTITUDE DETERMINES YOUR ALTITUDE

As a young professional, it is easy to think that your employer owes you training, great projects, lots of money, promotion, and so on. **Your employer owes you nothing**! Huh? "But what about a salary?" I hear you say. Well, my friend, if you are not contributing by bringing in work (in other words, money) or doing a great job on projects, you are not even owed that, as you will be on the retrenchment list. But what will set you apart and keep you off that list is your attitude.

3.1.1 Attitude, Attitude, Attitude and Working Hard

Have I said it enough? Your attitude will dictate your success in life. Many employers feel that they can teach you the skills required by the job, if they have to choose from a similar list of suitably qualified people. But one thing that cannot be taught is the right attitude. That you are not just there for the salary, but to make a difference, take the company/division forward, and be the best team member! Remember, you are being paid while you are building your experience/CV!

Similarly, being able to work in a cohesive team, understanding the value of synergy, and **putting the team before yourself**, those things are imperative to any employer. Note how the above does not require an individual winner, as life rarely does.

Finally, the best attitude is the one that recognizes that in the end, hard work will not only be recognized, but appreciated, while amassing the required skills and experience to become a first-class person in STEM. In business, you can be the cleverest person in the room, but if you can't work hard, work in synergy as a team, and bring in the money, you are useless to the business. Just work hard on the above and the rewards will come. Work on that which you can control: your attitude and bringing all you have to the team.

3.1.2 Asking for Work

Do anything. Do everything. No, you are not above doing any job. **ANY JOB!**

Once it is known that you are the willing, enthusiastic, up-for-anything employee, it will allow you not only to grow your skills, but give your manager confidence that the tough projects are ideally handled by you, as you enjoy the challenge and you deliver. And honestly, which project would you rather have on your CV? Smooth seas do not make for good sailors and all involved will appreciate that. How will you hone your skills without challenges?

Then, once a promotion presents itself, it will not go to the employee that easily navigated the easy task (not memorable at all), it will go to the employee that was challenged and came out on top through collaboration, hard work, and resilience (very memorable). Even if you failed at the task, the fact that you were willing to attempt it, and take inventive steps to overcome challenges, will ensure you stand out from the crowd.

Ask for the tough stuff; it will lead to personal growth and promotion!

3.1.3 Seeing Obstacles as Opportunities

If it were easy, everybody would do it. Yes, I know you are sick of hearing this, but no truer words were ever spoken. If it is difficult, chances are you are lucky enough to generate a job out of it and be able to take care of yourself and your family.

If you are faced with a difficult task at work, instead of getting angry with management, you should view this as an opportunity to show what you are made of. Rise to the challenge and just **try**! Fortune favors the brave! What favors those that don't try at all?

I want you to view everything as if it were your own entrepreneurial business. If there is something that is "hard," that creates an "opportunity" for you to do it instead of others, thus creating a job/business opportunity/advantage to promotion, why not try?

No one is going to think you are great for just doing your job. You get a salary for that. How will you make yourself memorable?

3.1.4 Getting Involved—The Story of a Bar Lady

No, it's not what you think. In my first job, after a period of time, buying stock for the company bar, packing fridges, and being bar lady every Friday afternoon at 4 pm fell to me. I took the task on with gusto! Why? Because we were a national company and the only way I could meet the national managers is when they came to the bar on a Friday after management and board meetings.

I was professional and friendly. When they later found that I was doing my bit on the business side and busting my butt, it was easier to talk about me, because they knew who "Michele" was—the friendly, professional bar lady assisting them on Fridays.

If you are the proverbial introvert, not participating in company activities (in other words, "I already give enough of my time during business hours"), then how do you expect to be promoted when no one knows who you are? When you are not willing to sacrifice personal time for company activities and be visual and friendly?

You are creating the image that you feel you are above the company. Don't. Participate and contribute to the company. It shows commitment and will give you a platform to show off your leadership skills. And your awesomeness, of course!

3.2 TAKING RESPONSIBILITY

3.2.1 Take Initiative

One of the most defining moments in my career was when I was supposed to design a treatment plant as part of a large team at my organization. My boss communicated to me that we would have a meeting where the different roles and responsibilities would be assigned.

Time came and went, yet the meeting just didn't take place. As I was new to the company and absolutely everyone in the team had more seniority than me, I was stumped. I felt new and did not want to point out that the lack of planning would come back to bite us. I also felt that as the lowest ranking person I could not arrange the meeting either. What I was feeling was the approaching speed of the oncoming deadline!

Eventually I went to the manager I felt most comfortable with and asked if I could start with some initial work just to better understand the problem. He agreed, and so I started putting the initial study together. Still the meeting with all the designers did not materialize. After completing the initial study, it was easy to move to the next step of the design. I roped in another newbie engineer and we started tackling the beast, so to speak. Still, no big meeting materialized. We just kept going until the design was done. And we made the deadline. That thing that clients value most.

Now look, it wasn't all just that easy. The other designers felt left out and were upset that the design was "given" to the less experienced ones (us). No good deed shall go unpunished—but on the back of this I was given a huge promotion, which to this day I still value and appreciate. The managers realized that we had taken initiative and delivered on time. That was the biggest lesson that anyone can get on taking the initiative.

If something needs to get done, take it on and do it! It's a big, beautiful opportunity!

3.2.2 Taking on Responsibility

Early in your career, the idea of taking on responsibility seems extremely daunting. You are still trying to figure out what you're doing and come to work every day feeling scared. All that you are hoping for is that one day soon you can do your job without fear.

Now, am I asking you to take on more? Yes, yes, I am, little penguin. No, I am not trying to torture you; I am trying to get you the best possible start in your career. You will not learn anything if you aren't being challenged. Smooth seas don't make for good sailors (again). I am saying the more you take on, the faster you will learn, and the faster the fear will leave you. Also, this will lead to faster promotions.

Not only that, but you will be known as the person who has **gumption** (look up this definition!), takes up a challenge, and has the right attitude, regardless of any typical hurdles such age, gender, or race. You will be judged on merit and nothing else because you deserve it!

Hard as it may seem to believe, having gumption is the exception, not the rule. If you want to stand out and be noticed, have gumption!

3.2.3 Learning to Make Decisions, Even Wrong Ones

Every day of my life, something happens to point out to me that fortune favors the brave. In this instance, being brave is being able to make a decision: using the knowledge at hand to make an impartial, possibly wrong decision, but a decision nonetheless.

The great thing is that your boss knows you are likely to make mistakes; it's human. It's also how you learn. But to sit around and not make a single decision is a road to disaster. You will be known as the spineless one who constantly needs to defer to management/someone else. Not on my watch, my little penguin!

I am not talking about leaving a question unanswered due to the fact that you need more information; I am talking about not wanting to take a decision out of fear, even if it may seem obvious. Gumption, people!

If you are not sure, ask, but please do not sit back in fear. Like a colleague of mine once said: Every day without a decision can cost millions, so have the gumption to do something about the problem!

3.2.4 Making My Boss's Life Easier

Your boss is also just human. Someone with dreams, ambitions, and the need for others to like them. They too want to be successful, appreciated, and recognized. Just like you.

There is a reason someone is your boss, respect that. Make their lives easier by thinking ahead and becoming indispensable. Insist on going to meetings with them so you can learn from them. The best part of this is also learning where they need help and providing it. And if it all goes horribly wrong, you learn how not to do things! As you know it's not easy to acknowledge that you are not coping and your function as a subordinate is to provide support to your boss. Support away!

Be proactive in taking on tasks that will lift some of the burden on your boss, but indicate you are doing so to free up time for your boss to find more business opportunities, otherwise they may feel threatened by you. That's another book.

3.2.5 Working on a Team

The best way to forge a good relationship with your colleagues is to work on the synergy that comes with being part of their team, but also know that you need not always take the credit. There is a famous saying that goes: "**If you don't need to take the credit, you can move mountains**." So, decide: do you want to move mountains and make this world a better place, or do you want to have credit for something much smaller because your pride will stifle the project?

This is also relevant when working in teams or across business units. Who cares who did what? The **TEAM** made it happen. The **TEAM** can make it happen again, or even bigger ... if you will let them!

3.2.6 Deadlines

Deadlines are non-negotiable. If you want to annoy your boss, colleagues, or clients, miss a deadline. Deadlines are usually time sensitive for a reason, be it funding, need, or politics, it doesn't matter. You set the deadline yourself (most probably). If you want to be famous for anything, be famous for the one always making their deadlines. **ALWAYS**. Enough said.

3.2.7 Becoming Aware of My Own Productivity

Do you waste a lot of time on email, social media, and chatting to colleagues? Take into consideration: what are you being paid for? Being productive. If you were spending your own money as a client on you, would you be happy with your productivity? Should the bosses have to choose to retrench or promote someone, would it be you (due to your current productivity)?

Why does it matter? It matters because there are other people out there like you, other companies like yours, and the only differentiating factor if you are delivering the same quality is: Yup, you guessed it, productivity, earning per hour for effort put in.

Don't be the one that's easy to get rid of.

3.2.8 Relationships with Clients

Clients are just people, too. People with families, financial pressures at home and at work, trying to deliver the best possible solutions for their stakeholders. They, too, are held accountable for the quality, budget, and time it takes you to deliver whatever project you need to do. They, too, just want to look good to their managers and advance up their career ladders.

Making your client look good is a win-win situation. Ultimately the people for whom you are delivering the service, for example the community, will win when the project is delivered on time, at a high quality and within budget. You win, the client wins, the community wins. Yay!

3.2.9 Making Promises

You Made a Promise. Keep It. If the production of the relevant deliverable is in your hands, you keep your promise. No excuses. **You made a promise**! Make sure that you are realistic about the deadlines you set for yourself and communicate with your client. It is **always** better to under-promise and then over-deliver, exceeding expectations!

However, if the production of the relevant deliverable is not in your hands, don't make a promise you can't keep. Do **NOT** promise your client that you can deliver within a certain time if you have not checked with your team. You will be setting your team, yourself, and your client up for failure.

3.2.10 Be Consistent

People don't want nice (promises of early delivery), they just want consistency. They want you to deliver a quality product on time, every time. **Consistently.**

Clients and co-workers also want you to be consistent in your behavior. They want to know who will be arriving at work tomorrow and what can be expected of them. They don't want a friendly, high-performer one day and a grumpy under-performer the next.

This applies to every part of your life: family, community, and at work. Just be consistent. In your behavior, in your delivery, and in your love. This creates a safe and secure environment for everyone involved.

3.3 PROFESSIONAL REGISTRATION IN STEM

3.3.1 Why You Should Do It

One of the most powerful yet most feared tools in our professions is professional registration. "Ag, no Michele! I don't wannaaaa ... and it's sooooo hard!" I hear you say. And that, my dear penguin, is exactly why you should do it!

"But there is so much prejudice and it's often unfair!" I hear you add. This is where I have to defend these professional bodies (for which I served on a committee). I want you to put your EI hat on: Imagine how difficult it is to hand out professional registration without upsetting someone, keeping standards high, and still complying with government legislation. Oh and of course, still being able to convince volunteers to give up their own precious time to do the assessments, yet still being able to pay administrative salaries so that the registrations can roll out. And on top of all this, try to stay relevant and try to preach benefits of being registered. Not so quick to criticize **now**, are you.

Yes, it's hard. It's supposed to be! It tells people you know what you are talking about. Professional registration implies **you made it in your profession** and you have not only the skills to implement what is required of you at a very high standard, but **MOST IMPORTANTLY**—you know when to say that you don't have the skill, experience, or knowledge to do what is asked of you. We all have different fields of specialization, mine being water and wastewater treatment process design. I am a professionally registered engineer. Under **NO** conditions should I ever be involved in the structural design of these treatment structures, unless I retrain myself in the structural design field. Don't be shy or afraid to say you cannot do something; it's not your field of expertise, and that the client should employ a specialist. Just as a geotechnical specialist will never tell me what process units I should use, I will never take a chance and ignore their recommendations. You know what you know ... don't mess around, as your registration will be stripped from you should you be negligent or found acting recklessly. Just don't go there. **EVER!**

Tied to your professional registration will inevitably be you signing some sort of code of ethics applicable to your field and legislation. This implies that every client will have a way to reach you should you act unethically, mess the client around, or try to rip them off financially. Just don't do it. **EVER!**

3.3.2 Tips on Registration

One of the best tips I can give you is to convey (depending on the specific requirements) **EXACTLY** what you did and how you were the hero. Think of it as a semi-interview: Why you should make it into a special club.

Use your wording carefully. Instead of saying "I was involved in..." say "I was responsible for ..." or "I specifically did...." Remember, the tea lady was to an extent also "involved" in your project. She made sure you stayed awake and made your deadline!

Be very specific. If you were the one interviewing you, what would you want to hear? Just a vague reference to a project, or something that would give you the confidence to give the stamp of approval, knowing your name is on that stamp? Give the people doing the assessment the confidence (and evidence) they need to give you this great honor. Otherwise, you just didn't earn it in their view, although in your head you know you did. So, coming back to registration being "unfair"—is it really unfair or are you just being unclear/lazy/arrogant in your communication? Why should these people put their heads on a block for you? Give them good reason to do so.

Another important tip is to write up your experience as you finish a project. One of the most difficult tasks is to try and remember what you did 5 or 10 years ago and then try to find your manager of back then to sign off on it. Even worse, what if you did not leave the previous company amicably? How are you going to get that experience signed off on? If you write as you go along, your report is ready by the time you have enough experience.

It's important to remember that registration is in your hands, not your mentor's/manager's. If you are spending years doing the same thing, you won't be eligible to register. It's up to you to find out what experience you need and to put pressure on your manager, ensuring you cover all the experience required in the least amount of years. If you don't, this can lead to a situation where you are not eligible for promotion, which may require registration.

Mentorship 4

4.1 IT'S A LONELY, LONELY ROAD TO WALK

Now that we have your creative juices flowing thinking about the many succulent morsels to be sampled from the buffet of possible careers/business opportunities, we need you to get some backup.

It doesn't matter if you are an entrepreneur going it on your own, or an intrapreneur trying to shoulder the responsibilities within a big corporation, at some time you will hit a proverbial (lonely) wall.

You feel intense loneliness in the burden that you have to bear and would do anything just to get some advice on how to handle the challenges facing you. Even if it is just someone listening and nodding.

This is where mentorship becomes extremely powerful. You will find companionship with someone who not only understands what you are going through, but who has broken through the challenges and come out stronger on the other side. With a touch of humor.

This person will be able to commiserate, have tips and tricks on what to look out for, and dole out wisdom and encouragement when doubt sets in.

No matter where you find yourself in life, you need a mentor. And even if it doesn't always feel this way, you have achieved in life and someone needs you, so also consider offering mentorship to others.

4.2 WHAT IS MENTORSHIP AND WHAT IS COACHING?

There are benefits to both mentorship and coaching and you need to decide what you need and when. There are a million definitions of both, of which I do not agree with all.

Seeing that this is **my** book (and you are welcome to reference others), I bring you my version: A mentor gets you through the long term, a coach gets you through the short term. In other words, a mentor can guide you into a long-term career choice and a coach can get you through a tough day. Like I said: my book—my version. Another way to look at it: The first gives you strategy, the other a standard operating procedure when in need. You, my penguin, will need both.

4.3 WHY A MENTORING RELATIONSHIP?

The reasons for mentees seeking out mentors vary greatly. Mentorship can be a formalized program in order for you to get some type of formal professional registration, or it can be very informal where you glean golden nuggets of wisdom. Either way, the opportunity exists whereby you don't have to make mistakes to gain wisdom; there are many people who can give you advice on the mistakes they made. If they say they didn't make any, they are lying!

It can refer to their life in your sector, being a female in a male-dominated world, being in a large corporate company, or being an entrepreneur, among others. You name it: someone's done it and probably messed up along the way. These people have also beaten down a path for you to flourish. Also remember that every time you mess up, it's knowledge you need to pass on. See—messing up is good for you! Otherwise, what will you be chatting about to your future mentee?

4.4 WHAT ARE THE RESPONSIBILITIES OF A MENTOR?

In truth, when a mentor commits to be a mentor, the biggest responsibility is support. It is important to understand the goal of the relationship and design a plan accordingly as set out above. It is also important to set clear boundaries to the relationship in order to manage expectations from both sides. Obviously, as a mentor you also have to be truthful about messing up.

4.5 WHAT ARE THE RESPONSIBILITIES OF A MENTEE?

The mentee has to be responsible for the required inputs to create a successful relationship and not just wait to be spoon-fed. The mentee needs to be very clear about the reason for needing the mentor, what inputs and outcomes are desired, and be realistic about the time that can be made available by the mentor. The mentee also needs to realize that conversations are private and confidential, and a mentor admitting mistakes to the mentee is meant to teach, not to provide fodder for gossiping.

4.6 PERSONALITIES AND POSSIBLE PITFALLS

When matching a mentor with a mentee, aligning personality types can be very important. A match between a big-picture person and a detail-oriented person, where one person may be expecting constant input while the other may be annoyed by being micromanaged, could lead to a breakdown in the relationship or a failure to build one in the first place.

Similarly, an extrovert trying to encourage an introvert to stand up for themselves and speak up may cause the introvert to retreat further.

A match between a spontaneous person calling when an idea strikes and an intense planner who does not appreciate short notice change of plans could be a huge source of frustration for both parties.

Ideally, personality types should be aligned as closely as possible, although, even if not, if these possible pitfalls can be managed, a successful relationship can still be enjoyed. This will depend on the specific relationship.

4.7 WHAT ARE THE RESPONSIBILITIES OF A COACH?

When coaching, the most important job you have up first is to get this young person through the day and give them enough courage to face another day. Then you need them to feel your passion to give them the courage to face one more day. From there, you need to make them see the difference they are making in this world so they can face **another** day! This part of the process is often overlooked.

From here you can **guide, not dictate**, what they should look into to be professionally registered. It still has to come from the young person to take on the job of getting registered.

4.8 BUILDING A CAREER PATH WITH YOUR MENTOR

The mentor/mentee relationship is built on absolute trust. Once you have abused this relationship, the taps dry up, either way. In your life, it will benefit you to be very clear on what you want in a mentor and to choose the correct one. I myself have gained enormous courage, support, guidance, humor, realism, and confidence from these relationships, both as a mentee and mentor.

You may have many relationships in your life, but this one will benefit you very much, if it is taken in the view of gratitude and enthusiasm.

This person has done it all and seen it all; they can open doors or guide you to another point of view. Give credit where credit is due and continue appreciating their input.

Your mentor wants you to succeed, I promise you. So ask, ask, ask, and ask again on what to do, where to go, how to upskill, and how to get to your destination.

Again, what you may need in your twenties will not apply in your thirties and later, so be patient with yourself, but give your mentor enough to work with to guide you.

4.9 IT'S A TWO-WAY STREET

Mentorship is one of the most powerful tools for success. However, one of my pet peeves WITH mentorship is that mentees may feel mentorship is owed to them. No one owes you anything (again). Mentorship is a two-way street and you should be honored that someone who most likely has no time to spare is making time to invest in your personal growth. Yup, I said that! You will only get out of the relationship what you put in and waiting for your mentor to spoon-feed you is hugely unfair.

That said, the relationship can be very rewarding and energizing for the mentor as well. Seeing someone succeed due to your input is both a fulfilling and humbling experience. I have often learned from my mentoring experiences. Getting to see something from another's perspective and trying to build their confidence is a challenging task with amazing rewards.

I encourage everyone to try to be in a mentoring and mentee role at all stages of your life. These needn't be formal agreements, just a knock on the door/cup of coffee/glass of wine when the going gets tough.

4.10 STANDING ON THE SHOULDERS OF GIANTS

One of the most comforting thoughts is the fact that no matter what trouble you run into in life, there is a pretty good chance that someone else has been through it. This is when mentorship means the most.

Whenever I get discouraged about being a female in a male-dominated world and the backlog in gender parity, I think of the women who fought for the right to study anything we want and vote for who we want, rights we tend to take for granted.

So if you feel yourself bumping that glass ceiling (and yes, it's very real), call on your mentor and those who have come before you.

The ability to build on the knowledge of others is very powerful; we don't need to reinvent the wheel, and we can use existing tools to solve problems. Another wonderful development has been the fact that all the knowledge so carefully gained by those who have come before us is freely available on the

Internet. I recall being at university and not having this privilege as it just didn't exist yet (yes, I am that old!). Those with the knowledge had all the power. Now anyone with a mobile phone has access to knowledge. Just remember though: Tough times and life experiences bring you wisdom.

4.11   LIFE IS A SCHOOL OF HARD KNOCKS

Whenever we try to encourage learners to enter a career in STEM, we try to provide a window into our world: the fact that what we do every day matters so much—we serve the community and make the world a better place.

Should we be successful in convincing them, often these learners start stumbling at the tertiary education level. Either the world differs vastly from school or they start to doubt that STEM is for them. Generally, this is still a level playing field and in the end, hard work takes them to graduation, regardless of gender.

What we owe them is a follow-up session at this time to say:

- Yes, it will be very hard—smooth seas do not make for good sailors.
- No, you are not owed anything just because you graduated—you owe yourself an amazing life.
- You need to pay your dues to get the experience and earn respect—use this time as a massive learning opportunity.
- Life is unfair, deal with it—learn the rules and play them.
- You will be treated differently as a woman/person of color/religious minority—advocate the advantages of diversity.
- Other people in your diversity group will make your life hard—be an example by pulling others up.
- Being a working mom will be even harder (the guilt will kill you)—but the rewards are unlimited (lots of hugs, kisses, holidays, and financial independence).
- It will get easier, we promise....

If we can set realistic expectations and provide support, we may have more young people and women staying in STEM.

Leadership 5

5.1 MANAGING PEOPLE

5.1.1 Inspirational Leadership (People Leave Managers)

If staff have left your team in the past, they left because of you. Yes, you. People don't leave jobs, they leave managers. Your team needs to feel inspired, supported, empowered, and encouraged by you. They need to see your strategic vision and adopt it as their own. They need to be able to rise to the occasion, not be smacked down for their ideas and efforts. Having staff too afraid to try anything, in fear of being ridiculed, or constantly being put down will create an atmosphere of stagnation and fear.

In the words of people who get divorced: "I know I have faults, but I don't need someone to point them out to me every day." Similarly, you may not have chosen all the staff you manage, but it's up to you to try and understand them and help them grow so that they exceed both your and their expectations. Kick those penguins in the water and give them wings to fly!

5.1.2 Be Professional

This is extremely important. By being professional you are leading by example, setting the tone of how others can expect to be treated and treat others. You are setting an example by dressing professionally, being on time, being prepared, and delivering quality work on time.

By not being professional you are not delivering a consistent message and indicating that inconsistent, unprofessional behavior will be tolerated or, even worse, rewarded.

5.1.3 Don't Get Personally Involved

This is a hard one. On the one hand, you will need to be able to have empathy and understanding. Your staff will have a pet that died, sick children, and tragedy in their lives. This must be handled fairly and consistently. However, there will be some who will try to take advantage of the relationship. Should this be allowed, fairness and consistency are lost. So getting personally involved will compromise your relationship and those who feel you are not personally involved with them will feel left out.

5.1.4 Treat Everyone Equally

Just like in any household, if you do not treat everyone equally, it will cause trouble and create resentment. Just be consistent.

5.1.5 Give Praise When It's Deserved

People do not only work for a salary, and expecting them to be grateful for one will not create hard workers. They can do the minimum for the same amount of money. (Many do!) Expecting a team to excel with no positive feedback is futile. I have heard managers remark that people need to be self-motivated and they don't believe in giving encouragement since that is what salaries are for.

By starving your team of praise, you will pave the way for your team to underperform or worse; your stars will move on to someone who they feel will appreciate them, instead of remaining with you.

So give praise when it is deserved (consistently—yeah I said it again), and help when help is needed in the form of positive critique.

5.1.6 Say Thank You

Not negotiable. Enough said.

5.1.7 Be Respectful

Treat **everyone** with respect. The old adage of treat others how you want to be treated applies. Another pet peeve of mine is when managers get irritated by younger/inexperienced staff. You too at some stage had to learn everything you know now. Allow others to make mistakes and do not treat them as if you can't wait for them to know it all. If they fail, it's because you are failing them! The reverse also applies. Respect the experience of others.

5.1.8 Don't Force Unrealistic Deadlines

Agreeing to unrealistic deadlines will not make you a good manager; it will only set you and your team up for failure. I have yet to find a client that fully appreciated and compensated me for killing myself and my team to fulfill an unrealistic deadline. Quality will be compromised and that is all the client will remember.

5.1.9 ALWAYS Take FULL Responsibility

As manager, you are ultimately responsible; not your subordinates. **YOU**. If something goes wrong, **YOU** apologize to the client, **YOU** find a way to fix it, and **YOU** ensure final delivery. **NEVER EVER** address your staff, partner, or sub-consultant/contractor in front of a client for non-performance. **YOU** did not adequately check quality or manage your subordinate/sub-consultant/contractor.

5.1.10 Working with Clients

5.1.10.1 Everyone has responsibilities

In addition to always taking full responsibility, it is important to understand that your client, too, has many responsibilities and similar challenges to you when they run a team. Putting yourself in their shoes will assist in your relationship with the client.

5.1.10.2 Bigger picture marketing

As a leader/manager you need to exercise and train your staff in big-picture marketing. This is when you leverage a smaller deliverable to use the opportunity to market another service or product offering in order to bring in more work or a larger project. Every opportunity is an opportunity for business development.

5.1.10.3 Everyone appreciates hard work and dedication

In approaching projects with dedication, not only will you increase your chance of more work with the client, you will be setting an example to junior members of your team. In other words, even though you are the leader/manager, you also work hard and don't just delegate all the work to others. You will earn the respect of others by grafting shoulder to shoulder, not just expecting them to work late.

5.1.10.4 Delivering realistic products (listen!)

This item is relevant to all aspects of life. We approach every situation with preconceived ideas and assumptions. One of the most important things in life is to be able to just stay quiet and let the client tell you what they need. I still struggle with this and try to work on this every day.

This will not only allow the client to be clear but allow for the development of ideas to improve the final product. By trying to preempt the needs of the client, you are missing an opportunity to solve their actual problems. This is ultimately what will keep them from coming back to you.

5.1.10.5 Going beyond the expected

Everyone wants to feel that the service being provided to them goes beyond their expectations. If you are able to achieve this within the available budget, do that extra bit and go beyond. Just don't allow yourself to be exploited.

5.1.10.6 Promoting the client

The client is just human, so if they are performing well, do take opportunities to promote them to their superiors. Everyone appreciates well-deserved praise, especially if it can assist their career. It has to be honest though.

5.2 DEVELOPING LEADERSHIP SKILLS

There are several ways of developing your own skills or the skills of those on your team, including:

- Public speaking
- Teaching classes
- Report writing
- Networking
- Working for free for a limited time period, then being selective
- Liaising with clients
- Voluntary associations
- Being recognized as a contributor to your field
- Bringing energy and positivity
- Bringing new ideas, innovation
- Being brave enough to be different
- Being authentic, honest
- Being human and understanding, generating loyalty
- Professional registration
- Management courses
- Conflict management
- Research
- Being a multiplier (read the book!)
- Develop a life outside work and family

Managerial skills training may include:

- Strategic thinking
- Leadership training
- Planning
- Business and project finance
- Risk and quality management
- Project management
- Staff management and motivation (also similar to human resources management)
- EI
- Many more which may, but not necessarily, culminate to an MBA or similar

Authority, Responsibility, and Ownership

6

6.1 WHAT IS AUTHORITY VERSUS RESPONSIBILITY?

Responsibility is typically a duty to provide a product or service to enable a profit for your company. Authority is being in the correct position in the hierarchy (having the applicable experience, tools, leadership backing, and title) at your company to deliver said product or service.

6.2 PITFALLS OF RESPONSIBILITY WITHOUT AUTHORITY

Why does it matter? Simple. If you are giving someone a huge amount of responsibility but not the authority to instruct others to assist them, you are setting your staff up for failure.

This is especially important when considering a project that needs input from multiple resources at various levels in the company; the person that has the responsibility to deliver on the project needs to have the relevant authority and the backing of management to instruct the whole team if the project is to be successful.

Remember, diversity creates better products and better profits, so managing diverse teams is an imperative!

6.3 HOW DO YOU FIT INTO YOUR COMPANY AND WHAT ARE YOUR RESPONSIBILITIES?

Obviously, only **you** can answer this question. You may have an appointment letter outlining your duties or the amount of business you need to bring in or key performance indicators (KPIs). You may be assessed against these KPIs in your personal appraisal which may affect your salary increase and bonus.

Are you clear about your role, its responsibilities, and its requirements? If not, why?

Are you supported by your management to execute the above? If not, why? What are you going to do about it, penguin?

6.4 HOW WILL YOUR RESPONSIBILITIES CHANGE AS YOU MOVE TO WHERE YOU WANT TO BE?

Earlier in the book, we envisioned your path. You need to be clear about whether you want to increase or decrease your level of responsibility. You need to communicate this and your reasons to your manager. If you don't, your manager may feel they are expending money and time trying to train you to take on more leadership or management responsibility while you may want to rather take on more business development, technical work, or research.

If the path you have in your mind is not communicated, it will breed resentment on both sides. You will only be productive and profitable as a resource when you are following your path with support, assistance, and the associated authority you need. But your superiors need to know what your plan is.

For those of you who feel the crush of responsibility, deadlines, and pressure—I promise it will get easier and easier every day. The responsibility, deadlines, and pressure will remain, but your ability to handle them will improve. I promise. Pinky promise!

6.5 WHAT IS BUSINESS OWNERSHIP?

Every country may have different legal descriptions for ownership, so it's best that you understand what ownership is in the country you are operating and the responsibility that goes with it. In the field of STEM, ownership is usually either corporate or small business ownership. Both have benefits and pitfalls, so research both before you approach this particular buffet.

6.6 WHAT TYPE OF OWNERSHIP STRUCTURES EXIST?

6.6.1 Public versus Private

Public ownership refers to stock being listed on the stock exchange. This stock can be bought or traded by anyone as opposed to private ownership, where the stock can only be bought by entities allowed to do so by the company in which the stock is held. This is an important aspect for you to consider in STEM. If you are part of a publicly listed company, keep in mind you will have shareholders who may have no idea what STEM is, but will expect a profit, regardless of you trying to change the world.

6.6.2 Financial versus Other Input

It may be that ownership in a new or existing company can be generated by contributing money, a specific skill or knowledge, a client base, intellectual property, or similar essential input. So, penguin, even if you may not have big bucks to invest, there may be something extraordinary about you that will add great value. Leverage this!

6.7 WHAT ARE THE BENEFITS OF OWNERSHIP?

When the company is doing well, shareholders will benefit through the payout of dividends. This is a financial reward/return on your investment. Other benefits as a shareholder include being able to vote/give input proportionally to your shareholding during shareholder meetings to ensure good governance at the company in which you have invested. Your hard work rewarded—yay!

6.8 WHAT ARE THE PITFALLS OF OWNERSHIP?

If the company is not doing well, the share value will decrease and you will lose out on your initial investment. Many people have lost their life savings in this way, reinforcing the need for shareholders to enforce good governance.

Even worse is, if you have private shareholding, you may be required to not take a salary or to pay in funds to the company to keep it going/pay salaries during tough times. In other words: Try to keep the company solvent and an ongoing concern. Yes, penguin, recessions happen!

6.9 WHAT ARE THE RESPONSIBILITIES OF OWNERSHIP?

So, to come to the point of this chapter, one of my pet peeves is when people are granted private shareholding, but these shareholders only want to enjoy the benefits of shareholding, also known as the dividends when things are going well. However, when the business is enduring tough times, they feel that the majority shareholders should be held responsible, and they do not want to give up their salary or pay in.

Your responsibility as a shareholder is to play your part when the going gets tough, scrutinize financial statements, and ensure that whatever management is up to is all above board and they are not only enriching themselves. You need to agree with current expenditure and point out matters if you feel the company may be able to cut costs on obvious items.

You need to protect your investment and ensure that the company and its management are ensuring the company is a going concern.

6.10 DO YOU DESERVE OWNERSHIP?

Are you willing to forgo your salary or pay in if it means tiding the company over? If not, why do you deserve the dividends?

When you are lucky enough to enjoy dividends, you need to ensure you have saved a certain amount to ensure that it will tide you over in tough times.

There is nothing as fulfilling to see that the harder you work, the bigger your financial return, and shareholding can do that for you. But be aware there will always be cycles in the economy and you need to be realistic about what your responsibilities are when the tide turns.

As in all things in life, you need to be fully committed in shareholding. To illustrate the difference between being involved and committed: when it comes to making bacon and eggs—the chicken laying the egg was involved but the pig was fully committed. Shareholding works the same way—you are all-in, my friend! Oink oink!

Communication 7

Throughout your life, you will hear again and again and again ... communication is the most important tool for a successful business (and a harmonious family life).

"Communication is key ... communication will make or break you" ... blah blah blah ... and yes, they are all very, very right. Sorry.

If you work on one thing this year, improve communication in your life ... effective, clear communication really is key. And watch your body language!

7.1 HOW TO COMMUNICATE WITH CLIENTS

One of the most important things to realize is that clients are just people, so they need to feel valued and they need to feel heard.

In initial meetings, you need to keep quiet and **LISTEN**! (Remember how I myself am also trying to work on this.) You need to communicate exactly what you heard or understood back to the client to ensure you understand the brief in order to understand the client's expectations. Not matching expectations with delivery will cause obvious unhappiness.

Once the expectations of the client are understood, it is important to constantly report back the relevant progress as required by the client. This will differ from client to client and be based on how well the client knows you. Remember to use these opportunities to communicate the slow advance of the required information if this is relevant. No client has ever said: "I wish you had told me what you needed later in the project." If you are awaiting information, make sure you have written evidence and bug them **OFTEN**! They will definitely not care about moving your deadlines unless pushed.

If possible, ensure progress meetings occur so that you can ensure you are on the correct path before spending time and resources doing something your client didn't want you to do. The client will not pay you for this! You weren't listening!

Instead of blowing your budget on the wrong thing, ensure you are able to break down the work in to smaller work products so that you don't go too far down the road before you realize you have it all wrong. Typically, an inception report will assist you in understanding the brief, but even these don't always cover all the needs/expectations.

Finally, if things go wrong, this is an especially important time to communicate as much as possible with your client. Often at this time, the client is most valuable and has a vested interest to assist you and bring the project back on track. Remember the client also wants a successful project and to look good. Often adversity will bring you closer to your client, and when success is achieved, it will seem so much sweeter to you both.

7.2 HOW TO COMMUNICATE WITH LEADERSHIP

Often, project managers mistakenly think that management don't want to hear from them. Well, if you are only approaching management when things go wrong, you will probably be right. The important thing is to constantly be in communication with management on the current status of a project, even if it is to say: "We are ok." Well, ok then, everything is ok. This gives management the reassurance to be able to give feedback to board level, even if it is to say: "No serious problems."

The worst thing you can do is to keep management in the dark when things start to go wrong. This is exactly the right time to let them know that a challenge has arisen and how you are trying to deal with it. Strange as it may seem, chances are huge that management has seen this problem before. Ideally, management will at this stage advise and, if possible, the challenge will be addressed.

The worst thing you can do is sit on the challenge until the outfall is irreversible and disastrous (and expensive). The saddest thing is to see a problem that could have easily been resolved up front be fought out in the courts, where only lawyers win. Not your case, just your money, and lots of it!

If management is constantly in the loop, they too can share tips and tricks and assist using their experience.

If things are going well, communicate the successes of your team to management! Give credit where it's due!

7.3 HOW TO COMMUNICATE WITH PEERS AND SUBORDINATES

The best way to describe this section is to ask you to imagine having a co-worker who needs your help, but you don't know what is expected of you, when it's required, and you don't have the information you need to carry out the task. You also need to be aware of any changes or updates. How are you supposed to assist when you don't know what's going on?

Put yourself in their shoes and it will become very obvious what is required of you as the project leader. **Open, consistent, and clear communication**. Request that the person repeats what they think you require of them so that you are assured you are understood.

Always remain friendly, respectful, and professional and make it clear that you are grateful for their valuable contribution. Clear, constant communication.

If you are not consistent and you are constantly changing your requirements, the person will put your requirements on the back burner until you have sorted yourself out. You will lose respect and relevance. Your deadlines will become guidelines and their input will be mediocre as they expect you to change your requirements anyway. Don't mess people around, no matter their position in the company hierarchy; it will blow up in your face. Ever heard of "Cry wolf?"

7.4 BENEFITS OF FREQUENT COMMUNICATION OR "CHECKING IN"

If people are used to you "checking in," they will understand that it is you trying to ensure that they have all they need to be productive and that you are not micromanaging them. Try to use the words: "What do you need from me?" It will reassure them that you are there for them and that their productivity is now in their hands. If they are not performing, they have no excuses.

It will also be of benefit to you, as they will be expecting you at a certain time and will collate their queries for when you will check in, instead of bombarding you on and off with less urgent queries. If you do receive something outside these check-in times, you know it's urgent and needs your immediate attention.

7.5 PITFALLS OF COMMUNICATION AND TECHNOLOGY

Here are my tips gleaned from my many, **MANY** mistakes.

7.5.1 Emails

- People do not read long emails. People often miss very pertinent information the longer an email gets. Highlight important deadlines/information in red, yellow, bold, or larger font.
- **An email CANNOT be deemed urgent! If it's urgent, CALL them.** People can get stuck in meetings and it is frowned upon not to give them your full attention by screening emails.
- People ignore repetitive emails with the same information. Make it relevant and current.
- Don't spam people's in-boxes by overloading them with emails; it is a sure way to be ignored.
- Be concise and clear. If required, try to add details in attachments but be aware these may be ignored.
- Be aware that being able to read emails on your phone can often have the opposite desired effect by inadvertently losing emails, or the view screen may be poor. Ask me, I have lost many emails this way... gmf! So, if someone says they didn't get your email or did not see something, it is often true (think junk mail also).
- So, follow-up promptly on your emails, assuming it may have gotten lost so as not to compromise deadlines. The onus is still on you until you get a response.

7.5.2 Phone Calls and WhatsApp

- Phone calls are used by professionals, WhatsApp by friends (but it has its uses).
- If you are communicating internationally, call with WhatsApp if you have free access to data—it will expand your network and break down borders at no cost.
- If there is something too important or delicate to discuss in detail by email, call first to set the scene and let someone know to expect

the email and your requirements of them, including the timeframe. It is amazing how you can maintain good communication and good relationships by not dropping bombshells and preparing people with a simple phone call.

- Follow-up on any important telephone discussion or meeting with an email to ensure that you understood correctly and include deliverables and expected timelines. Then deliver on those. This also leaves a virtual paper trail should you need to defend your position if things go wrong.
- Do not unnecessarily add people to WhatsApp groups; they will get annoyed and leave the group anyway.

Marketing and Selling 8

There will always be someone cheaper than you and there will always be someone better than you (and more attractive of course). So, how do you set yourself apart? Why should someone use your service/product?

Thinking out of the box and value-added service. Um, ok then … but don't forget to also give them what they want. What is the problem that you are trying to solve?

And once you have a client, how do you keep them? It takes so much more effort to get a new client than to keep an existing one—yikes! Better to keep them then, don't you think?

Did you know that your client understands that you're human and make mistakes? If you are open and honest, they will give you the opportunity to fix it. However, your biggest mistake in business will be an attitude of indifference toward the customer by you or any employee associated with your project/company. **For this, a client will walk away!** Don't believe me? How did you feel when you were getting the run around from a call center? Made to feel as if you are just an annoying fly bugging them. Bang—indifference!

Can you believe that even though you did nothing wrong, you still lost a client! Just because you didn't make them feel valued? All you did was act indifferent. Stop it! Now! Make your clients feel special, they deserve it! Or you don't deserve their money!

8.1 EXTERNAL

EVERYTHING YOU DO AND SAY represents you and your company. Your weekend activities, your conduct at all times, and especially everything you do on social media. This I cannot stress enough! Although not always possible, try and keep your friends on Facebook and business contacts on LinkedIn. Try not to tweet at all, even when you mean well. I'm sorry; this is for your own protection. No good deed will go unpunished.

The client is always right! Focus on the solution rather than the blame. No client will ever take the blame, they are just human. They don't want to get in trouble either. But they will choose to work with you again, if you have their back and deliver them from disgrace, gracefully.

Treat clients as if it's your own company on the line, because actually, it's **your reputation** on the line. Clients don't remember your company, they remember you. Good or bad.

Be the preferred service provider by being

* Friendly
* Professional
* Dependable
* Personable (ask about their interests, you will be shocked!)
* Going the extra mile—but don't become a doormat

8.2 INTERNAL

Selling yourself in your own company will be one of the best ways to get ahead. But you won't be getting anywhere if no one knows who you are.

Be visible: Get involved, be it family days, company challenges, company events. You will only be able to convince all directors that you should get that promotion if they have laid eyes on you.

Be dependable: Now that you are visible, remember marketing internally is only successful if you are found to be dependable. This will allude to bad marketing if you cannot be depended on in these informal occasions.

Be professional: At all times, in everything you do. Your manager needs to know that if they take you to a meeting that you will conduct yourself in a professional manner. Be on time, well groomed, and prepared.

Be consistent: Everyone wants to know the person arriving tomorrow will be in line with their expectations. If you are not being consistent, in your attitude, appearance, professionalism, or the quality of your work product, your managers will resist taking you to clients as they will fear what your response will be on the day. They will not give you important projects as they will not know what to expect. Be consistent, be dependable. Always.

Be friendly: This may seem obvious but it is highly overlooked. Consider the people you approach on a daily basis. If you are in a queue and get to choose to approach the counter of the friendly assistant or the grumpy one, you can't help but approach someone friendly. Powerful stuff!

Now, consider your behavior at work. If you want to be chosen to be part of the top team, taking on the top projects, and your competition has the same credentials as you, why should you be chosen?

If your client has to choose between you and your competition who has the same credentials as you, why should you be chosen?

Simple: Because you are the **friendly, amicable person**. It's easy to be around you and you are easy to talk to. You choose to put the team first and you know that if you don't care who takes the credit, you can move mountains.

8.3 NETWORKING

One of the most powerful tools in business development is the art of networking. This is a book all on its own and my recommendation would be to take a course on this subject. Here are some tips:

- Anyone can be a possible client/contact, so stay professional at all times.
- Help others connect to your network, as they will do the same for you.
- Read the potential contact's body language; they may not be in the mood for you.
- Work on the relationship before pushing any business opportunities; they need to feel that they can trust you before vetting you.
- Work on a joined solution with colleagues so that you can use your and their contacts for the combined solution.

The other tip is not to push your business onto friends, as this is a surefire way to alienate your friends.

8.4 PERSONAL BRANDING

It doesn't matter where you work, people will remember you and your work ethic. They will follow you from employer to employer. Work on your personal branding, as your name will follow you in your industry no matter where you go.

In order to uphold personal branding, the following is suggested:

• Be friendly, always, **ALWAYS!**
• Make sure you excel at your job; hard work is always appreciated.
• If a project allows, see how you can incorporate training and try to inspire others.
• Do more than just your job: Volunteer with associations/committees, and expand your network.
• Connect with others in your industry to find out how you can give back.
• Be your awesome self, **ALWAYS!**

Intrapreneurship 9

9.1 SEE YOUR UNIT AS A BUSINESS

The best way to navigate your way as an intrapreneur is to imagine your unit as a stand-alone business. This business needs to have certain resources and systems in place, needs to bring in a certain amount of business to cover overheads and yield a desired profit, and needs to deliver a quality product in a certain amount of time, within an agreed budget with the client. It is also important to remember that **cashflow is king**, and although you may have the best order book, without money coming in, your unit is not a viable business. Oh, and most of all, profitability is key!

9.2 WHAT IS YOUR VALUE ADD?

In this day and age, you need to understand that although you are part of a business, you need to find something that sets you apart from your competitors, should you have been a stand-alone business. What is your value add? How are you adding synergy to the rest of the business or to your field of expertise?

Ultimately, if times get tough, you need to be able to answer the question: "Why should your unit remain in existence?" or "Should your whole unit be retrenched/liquidated?"

9.3 LEVERAGING OTHER BUSINESS UNITS

One of the most daunting aspects of being part of a larger company is being able to generate enough business and consequently enough profits to sustain the business at large. This is a big monster to feed with large volumes of salaries that have to be paid on a monthly basis.

Instead of seeing the units as individual service providers, it is imperative that the units be seen as possible business partners that can generate more revenue than would have been generated as separate units due to the phenomena called synergy (yes, penguin, again).

As a whole, more products can be developed, more services can be offered, and a larger knowledge base can be drawn from. The teams can learn from each other and additional aspects, incorporating more than services previously offered by all the units involved can be included, such as artificial intelligence, data mining, and virtual reality.

Not only this, but there are more contacts that can be mined for the additional products and new service offerings made available.

Finally, all business units can then market for all the other business units, spreading the net far wider than would have initially been the case.

By working as a larger unit and not competing against each other, more business opportunities can be developed to grow all the units involved. The concept of "healthy competition between business units" more often than not results in unsustainable backstabbing, instead of a combined sustainable growth achieved by the efforts of all parties involved. This fact is important in all aspects of life, by the way. Synergy!

Entrepreneurship 10

10.1 FINDING OPPORTUNITIES

The most important role of an entrepreneur is finding business opportunities. Being able to see an opportunity in challenges and being aware that if something is difficult, not many will do it—therein lies a business opportunity.

Entrepreneurs are often people who see opportunities everywhere and will struggle to focus on just one thing. This is not necessarily a bad thing as ideas can be combined, but it can be a challenge if you are unable to focus enough to bring the business idea together.

10.2 YOUR TRUMP CARD IS BEING FLEXIBLE

You have no red tape, you lucky bugger, so why not be flexible in meeting the needs of client! Be agile and adjust to the market, the need, the budget!

10.3 BANKABILITY OF IDEAS

It doesn't matter how great your idea or opportunity is, it needs to be bankable. You need substantial startup capital and will need to sustain your income for some time while you get your business off the ground. You need to create a business plan setting out the purpose of the business, projected income, and cashflow as well as a thorough SWOT (strengths, weaknesses, opportunities, and threats) analysis.

Ideally, you need to have a project or opportunity already in the bag with which to get the business off the ground. But even then it will be hard. Very hard! But this is what being an entrepreneur is all about.

10.4 USING EVERY OPPORTUNITY TO PUNT YOUR SERVICES

Being an entrepreneur means being brave and using every possible opportunity to market and network. It means being aggressive with all available leads as if your bread and butter depend on it. It means using every possible contact and avenue to generate buzz around yourself and your skills, product offering, and excellent service. This is your time to **SELL**!

10.5 YOU'RE NOT ABOVE ANYTHING

You need to be able to do anything and you need to be able to do everything. Although initially this may seem like an insurmountable challenge, this is the time you will learn the most about business and yourself.

You need to familiarize yourself with the latest legislation regarding commercial and tax laws and understand governmental requirements.

You need to understand how you can limit risks such as the use of quality management systems and indemnity insurance and understand all the jargon that comes with this.

10.6 ETHICS

It takes a lifetime to build your reputation, but it only takes one stupid mistake/taking a chance/desperate measure to kill your reputation.

If you are wondering if what you are doing is unethical, consider if it were front-page news. If what you did will embarrass or humiliate your family/employees/business partners, just don't do it. There is no time limit on corruption charges.

10.7 TIPS TO GET GOING

Did I tell you about being friendly? Don't forget the client's need to want to be around you. Your clients need to feel that you put them first, above all others. They are taking a chance on you and you need to assure them their trust is founded.

Do more, and more, and more. You need to make the client feel that you will go above and beyond what is required and give them a reason to continue using you.

You will probably need to take a bit of abuse at the beginning; small companies often bear the brunt of this as clients know you cannot afford to upset them.

You will need to do some stuff for free ... yikes! But consider that anything done for free can also be viewed as marketing, having an opportunity to have face time with a respected client, and being able to show the client firsthand how excellent you are!

But remember, cashflow is king and your cashflow will need to be able to cater for all of the above.

Innovation 11

How will you differentiate your service offering (in your own company or as an entrepreneur)? Without knowing you (the awesome, friendly, professional go-getter), how will the client know to choose you?

11.1 INNOVATION IN SERVICE

11.1.1 Value-Added Service

If a client can get the service you are offering from many other service providers, why should they choose you?

What advantage can you deliver to a client that trumps the value offering by others? The client wants to look good, seem progressive—what can you bring? Often the easiest route is to do things the way they have always been done. **THINK**! If you didn't know how it's always been done and you are spending your own money and not the client's, how would you change the offering? This will give you a competitive advantage.

Deliver more than expected. Go beyond the norm of what others would have delivered. Don't set a precedent for abuse, just set a precedent where it is easier for the client to choose you as you are the one that delivers.

11.1.2 Fit for Purpose

In order to deliver on the latest innovation, you need constant training. To stand above and be a success you need to be aware of the latest local and international trends in your business. Is your offering outdated? Is your large corporate company archaic in its thinking?

Will your company be able to incorporate the fast-paced developments in artificial intelligence, virtual, or augmented reality, yet still be able to adapt to local conditions in being robust and fit for purpose?

You need to be sure that you understand the needs of the clients and apply appropriate technologies that are easy to operate and maintain.

The best way to be up-to-date with the relevant advances in your field is to stay up-to-date with conferences, publications, and one of the most powerful tools of our time: the Internet!

11.1.3 Intellectual Property

As you become more specialized in your field, you may have the joy of developing intellectual property (IP). This can include certain innovative procedures, processes, or software. Is there something running around in your head that can be monetized? Will you be first to market it? Is your IP protected by, for example, a patent? Do you understand the applicable IP laws of your country and internationally? Is it adaptable to your client's needs? Innovative IP is a must that should be encouraged as much as possible. Not only will it set you apart from your peers, it will provide annuity income.

11.2 HOW TO FOSTER INNOVATIVE THINKING IN YOUR TEAM

One of the best lessons I recently learned is that I was brainstorming the wrong way all my life. I assumed you threw out ideas, they were judged for merit, and then implemented.

One of the pitfalls of this model is when someone on your team proposes new ideas that are being shot down all the time, they will retreat and stop contributing. It will also not allow the combination of individual ideas.

The best way is to let the team throw down ideas over a period of time, without any ideas being shot down. It can even be an anonymous board in a communal area, which lends itself to the team being in repetitive contact with the ideas.

From here, the ideas can be combined to bring forth even better, more practical ideas, using two or more of the ideas on the ideas board.

Only now will the anonymous ideas be judged for practicality by the team, as a whole. No judgment of individual ideas, just a communal learning session on the realities and practicalities of implementing ideas or even a SWOT session. Implementation only follows after this. The benefit is team buy-in and immense synergy. Done and dusted.

11.3 SOME IDEAS THAT WILL NEED YOUR CONSIDERATION IN INNOVATION

Here are some considerations for getting the creative juices flowing:

- Fourth Industrial Revolution
- Global WiFi—yes, it will happen
- Access to mobile phones in Africa
- eWallet
- Artificial intelligence
- Virtual or augmented reality
- 3D printing
- Business analytics
- Blockchain technology
- Water, energy, and food security
- Online shopping surpassing physical shopping in 2019

Financial Intelligence 12

If ever there was the proverbial "knowledge is power" truism, none is truer than understanding finance. **All aspects of it**. The purpose of this chapter is not a crash course in finance, but to assist with tips, tricks, and typical pitfalls.

12.1 PERSONAL FINANCE

You will not be able to master business finance if you are not in control of your personal finances. "Noooooooooo!" I hear you say. Yes, my friend. Oh yes. Stop buying shoes!

"But Michele!" I hear you say. "That is my partner/husband/financial manager's problem!" No, it's not. "But they are good at it and I am not!" I don't care. How can you run a business/department when you run away from your own responsibilities/finances? I will not promote you to run a department if you are clueless on personal finance! Time to face the music.

12.1.1 Know What Your Expenses Are—And What They Will Be in the Future

- You need to know what each personal debit order or transfer is for. You need to know why each cent leaves your account.
- Is this debit order still relevant? Are you paying for something you no longer need, such as cover in case of not making a finance cover payment on something you already paid off?
- Can you reduce any of the payments, such as insurance on your aging car?

- Are you aware of the portion of money you are spending on necessary debt (home loan) versus unnecessary debt, such as from store cards?
- Are you aware of the interest rates you are currently paying on each account you have (credits cards, overdraft, and personal loans)? Can you consolidate your debt into a product with the lowest interest rate?
- Are you paying off your debt as soon as possible?
- Are you living beyond your means?
- Do you understand the stock market and the effects on your pension?

12.1.2 Know What Your Income Is—And What It Will Be in the Future

You need to assume that you may not get a bonus or an increase. Or you may be in line for a salary cut. Will you still be able to cover your future expenses keeping in mind the annual increases involved? If you are a shareholder, consider that you may not receive a salary at all while times are tough or you may have to pay in.

- Do you have a large enough buffer in your income to accommodate such an above-inflation increase?
- If not, is there a way you can expand on your income, such as income via hobbies or other interests?
- Are you paying enough for your retirement and life/disability insurance? These will be future incomes.
- Are you saving enough for a rainy day in the event of a salary cut, salary pause, or retrenchment?
- If you have enough savings for a rainy day, are you placing your money in the right places to maximize your investments (in other words, not only in normal banking products, which do not keep up with inflation—not advised)?
- Do you understand the stock market (hopefully), and are all your eggs in one basket or evenly spread across various markets (finance, technology, mining, etc., or stocks, cash, and bonds—highly recommended)?
- Are you an active or passive investor (read up—this may surprise you!)?

If you are not sure what to do, consult a financial advisor. Investing can be fun, exciting, and empowering. Knowing you chose a stock based on good financial analyses and see it grow can start your financial emancipation.

12.2 BUSINESS FINANCE

12.2.1 As a Young Person

Every second you spend at work should be spent covering your salary or more and trying to find future business to enable you to continue working at that particular company. Once you can't cover your salary you become a liability and as such you go on the possible retrenchment list. Yes, I am trying to scare you!

So next time you sit and waste valuable company time and resources by surfing the Internet or trolling on Facebook (I see you!), consider the fact that you are not spending every available moment setting yourself apart from your peers to bring in more income or become productive. You are giving management reasons to forget you and you are creating an opportunity to move up the list of possible retrenchments.

Every moment at work should be spent finding work, client liaison, delivering results, improving the quality of work, or building work relationships.

It should be spent setting yourself apart; being productive, memorable, and profitable; and being at the top of mind for promotion.

12.2.2 As a Business Owner

Some of the items listed below may not seem like financial items, but ultimately, they turn out to be.

12.2.2.1 Cashflow

Cashflow, cashflow, cashflow. You can have the most amazing business in the world, but if you are having severe cashflow problems, your business is at risk. Stop giving interest-free loans, which is what outstanding debt actually is (also known as clients that do not pay).

It's tough: How do you handle a client who owes you money when you want repeat business? I know it's not easy but appeal to their human side.

12.2.2.2 Death and taxes

Don't ever skip a tax payment. **EVER**. Enough said ... ever! They know who you are and the interest will kill you. If not that the prison sentence will. Ok, so Martha Stewart rocked it, but she is the exception.

12.2.2.3 Insurance

You can deliver the perfect project, and out of the blue some strange circumstance will throw a crippling problem your way. Lately, geotechnical studies, which are normally not even part of your main offering of consultants, or that expensive, are the culprit. Never try to get a discount on the geotechnical investigation!

12.2.2.4 Business development

Business development must be done at all times to keep the pipeline full; the monster remains hungry and those salaries won't pay themselves. Even if you have too much work, don't skimp on this.

12.2.2.5 Your reputation

Again, it takes decades to build your reputation, but one bad decision to ruin it. Don't ever rip off a client; no amount of profit is worth it. Your reputation will precede you.

12.2.2.6 Business partners

Wrong partners or clients can tank your business. This is not a charity; salaries and overheads need to be paid!

Even worse, partnering with someone dodgy hoping to access their dodgier connections is looking for trouble. Choose wisely, even if it means sacrifices.

12.2.2.7 Integrity

Don't do anything that you won't mind being front-page news (again). Corruption is a choice.

Risk and Quality Management 13

13.1 WHAT ARE THE TYPICAL RISKS FACING YOUR COMPANY?

13.1.1 Financial

The biggest risk for your company is not having enough cashflow for your company to keep going. All the other risks eventually speak to this. No money, no business.

Looking back at previous chapters, by now you should understand income and expenses. Unfortunately, life is not even as nice as that. Recessions happen and you need to use savings to make it through. But the question is: Until when? Should I add other divisions (also known as diversify) where there is a need, or should I change geographical location?

13.1.2 Liabilities

Always keep your eye on the number of liabilities in your company/division. How much do you owe sub-contractors and who do you owe it to? Please do **NOT** skimp on paying taxes! They will find you! Get a financial advisor to assist you; they will try to assist you to minimize expenses and taxes.

13.1.3 Insurances

You cannot go through life without insurance, no matter what your business is—cover your back. Make sure it covers relevant issues and make sure it is enough. **AND MOST IMPORTANT**—make sure you know what is covered! Whether you are a young person, manager, or CEO, you need to know!

13.1.4 What Are the Typical Risks Included in Your Projects?

Budget: The project needs to remain profitable, otherwise you are on the retrenchment list.

Resources: You need to be assured your staff can do the job and that you have enough of them on board.

Time: My pet peeve—rather under-promise and over-deliver. Why nail yourself on time—YOU get asked for the initial program, why be unrealistic about it? No one is going to love you for it!

Contract law: This depends on where you are and what you are specializing in. But, as always, try to understand what you are getting yourself into before signing **ANYTHING**! Ask a lawyer if you don't know! Making use of a lawyer at the contract stage may seem expensive, but it could save orders of magnitude in costly litigation later.

Always watch what you say or promise! If you keep on saying: "No worries: we will sort it out," then consider that your insurance may not cover these things you are promising!

13.2 QUALITY MANAGEMENT PITFALLS

All quality management systems (QMS) can be bypassed or be a paper exercise. Worse is when this duty is delegated as punishment. Any QMS is expensive if not used properly. But it can also be expensive to the company if there is too much paperwork that may not be relevant or fail to address trivial issues.

Make sure you understand your QMS system as both a young person and business owner. Make sure that the relevant business processes in your QMS is focused on consistently meeting the requirements of your customer and identifying and mitigating relevant risks. Make sure you know what you are paying for and take ownership of it. The auditor audits the paperwork in front of them. In other words: are you doing what you said you would do? It does not prescribe **WHAT** to do. So, make sure the **WHAT** is clear, relevant, sustainable, and fulfills the mandate of what a QMS is supposed to do.

Fortunately, the newer ISO 9001 QMS systems are more risk based, helping you as a business owner to address these.

Sustainability 14

14.1 WHAT IS SUSTAINABILITY?

There are a million definitions out there. I encourage you to find your own. Ultimately, you need to create an environment where you can do what you love every day, but not to the detriment of something else. Making life a bit better, remember?

14.2 WHY IS SUSTAINABILITY IMPORTANT FOR A COMPANY?

Sustainable development has now moved into the mainstream of political and business thought, because future availability of energy, water, and non-renewable materials is at great risk. Including sustainability in design is no longer a nice thing to have but imperative for the future of the planet and the future of our industry.

This translates to the design of resilient infrastructure that will be able to cope with the extreme conditions we can now expect on a regular basis due to climate change.

In all challenges, there exist new business opportunities. Being able to recognize those opportunities and remaining unafraid will set you apart from the rest. It will create a competitive edge, setting you up for success. Instead of being in denial about climate change, rather think how you can make a difference.

Sustainability in a company is not only about the changing world, but also ongoing relevance of the company and competitiveness in a changing market and environment. You have to be able to adapt to, for example, ongoing digitalization, the need for increased productivity, or changes in procurement styles by clients.

14.3 WHY IS SUSTAINABILITY IMPORTANT FOR A PROJECT?

14.3.1 Resilient Infrastructure

Being able to foresee weak or critical points in infrastructure will set you up as a pioneer in the design of resilient infrastructure.

The design of resilient infrastructure is not rocket science and it need not cost exorbitant amounts of money. Often it is working in partnership with the private sector in recognizing advanced products already available that can be used in innovative ways, such as new cement products or pathways that can collect stormwater for water reuse.

14.3.2 Lifecycle Costing

One of the most important tools in sustainable design is the concept of lifecycle costing. Often initial infrastructure cost is the only consideration in design for clients as they are forced to use price/cost-based selection. This will then often equate to higher operation and maintenance costs during the life of said infrastructure, particularly energy costs. Another example is the purchase of cheaper options that do not last the typical lifetime expected by the unit, such as pumps. By keeping the cost of the lifecycle of the infrastructure in mind, the total project becomes more sustainable, as many clients often cannot afford the high operation and maintenance costs involved after low-quality infrastructure construction has been completed.

14.3.3 Environmental Impact Assessment

Luckily, in most countries, it is a requirement by law to conduct an environmental impact assessment to ensure no permanent damage is done to the surrounding environment by a planned project. This encourages the sustainable life of animal and plant species affected by a project. It also looks at affected water bodies and geotechnical issues that may arise, such as dolomitic conditions.

14.3.4 Social Issues

An aspect often forgotten, but which may have a huge financial impact, is social acceptance. Before large amounts of money can be spent on an expensive project, affected parties need to be consulted. This is often included in the environmental impact assessment as a public participation meeting.

It must be stressed though that although your project may have passed such a meeting, you still need to ensure buy-in from future users of the infrastructure. In such a case, your public client is not the ultimate client, but the user will be. Traveling into the community or using specialists to do so will focus exactly on what the community needs are, which will guide you in the effective use of available funding. The worst-case scenario of non-sustainability is using taxpayers' money for infrastructure that is ultimately rejected/abandoned by the proposed users.

14.4 HOW CAN YOU MAKE A DIFFERENCE?

By using EI (placing yourself in the user's shoes), by doing proper market research into the needs required, and by researching different future scenarios, you will already be one step farther in creating a sustained user base for your infrastructure/service offering/product.

Taking that extra step and by using common sense in what could be affected by further consequences of climate change, you leave your competitors in their denial.

The best model for this is, again, scenario planning and being ready with a feasible backup plan for each scenario. Even for the worst case, there is an answer in STEM. Exciting, isn't it?

Attracting, Developing, and Retaining Women Engineers and Scientists

15

15.1 FACTORS THAT IMPACT ENCOURAGEMENT INTO THE FIELD OF ENGINEERING

15.1.1 Initial Factors

15.1.1.1 Wow factor

The greater the "wow factor" at a young age, the greater the interest and engagement, which is linked to the power of engineering to make a positive change. When you have an opportunity to tell young learners how engineering contributes to society and your experiences in engineering, do so! It's so easy for young people to take roads, bridges, water, sanitation, electricity, and multistory building (just to name a few) for granted. Remind them that it may seem like magic to open a tap and get hot water, but it's engineers that make that magic happen!

15.1.1.2 Helping society

Although a generalization, I have found that I have made more of an impact with prospective female engineers when engineering is perceived as helping and benefiting those in need or society, in general. I have found many young people want to make a difference, which engineering provides for.

15.1.1.3 Income

Although another generalization, I have found that I have made more of an impact with prospective male engineers when engineering is perceived as a career with an above-average income. Initially this is for the achievement of status (possibly), and later to provide comfort for themselves and their families. However, in later years, I have noticed their interest again changes to wanting to make a difference. I agree I am generalizing, but I am trying to help should you have limited time to make an impactful speech.

15.1.2 Building Awareness through Career Development

Other than the informal manner stated above, there are luckily many formalized systems that can have a great impact in building awareness:

- Guidance counselors
- Career centers and associated competitions
- Job shadowing
- Sponsorships

15.1.3 The Use of Media

Very successful media campaigns have been run across the world to encourage the study of engineering, most notably in Australia, which ran formalized advertisements on television. Other options are:

- The use of social media to create a buzz around the opening of large infrastructure projects
- Profiling successful projects that are making a difference in society on all possible platforms
- Profiling successful engineers and their career paths

15.2 THE VERY REAL CHALLENGES FACING YOUNG WOMEN IN STEM

15.2.1 Waiting for Someone to Save You

From a young age, girls are indoctrinated to "wait for her prince to come save her." In a patriarchal society, this can often be reinforced throughout school years into further studies. By the time you start working in your STEM career, it can be easy to still be waiting for permission/help from male seniors.

It's very important to teach these young women that there is no one "coming to save them." Additionally, the manager does not "owe" them anything and it is up to the young women to **save themselves**. They have to be the hero in their own story. They must not hand over their future to anyone and make decisions based on pressure from others. Young women need to actively seek their passion, request appropriate training or qualifications that will fulfill their passion, and not leave training or the future direction of their careers to their managers. In all fairness, the managers will plug a hole where the need is and do what is best for the company at that time to fulfill the current obligations.

15.2.2 Family/Community/Societal Obligations (Perceived or Real)

There is no bigger guilt than the guilt that can be heaped on someone by their family. These can be extended to perceived obligations by the community or what we think society expects of us. Once children come into the mix and mommy guilt is added on top of this, a young mother with a hectic work deadline can be made to feel that although she feels split into all directions, she is simply not managing to satisfy anyone; and a feeling of hopelessness may prevail. In the words of Gerard Depardieu in the movie *My Father the Hero*: It doesn't matter "where" she is, she is just not there. Somehow she is just not "there" for anyone.

15.2.3 Historical Roles

It doesn't matter that I have been an engineer with a doctorate for decades; when I am in a meeting and I pour myself a coffee at the coffee station, I feel it is expected of me to offer to make everyone else coffee, even if I am not in my office boardroom. Which is nonsense! This is only one example. There are other historic roles attached to culture, religion, and so on, that have and will always affect women.

15.2.4 Fighting the Glass Ceiling

I've heard young male engineers say, "What glass ceiling?" or even worse "It's not in a woman's nature to be a leader/CEO." With the same logic, women can say it's not in men's nature to cook, which is obviously not true. I dare you to say that to a Michelin Chef's face.

15.2.5 Much Older Subordinates

It's difficult to project manage people your parents' age, but in STEM this is a common reality. This is where good company management is critical in giving young managers both the responsibility and authority to carry out the required tasks. But even then, it still remains a challenge that needs to be handled with respect from both sides.

15.2.6 Much Older Contractors

Doing any type of contract management is difficult, but being faced with contractors that keep on pointing out your lack of experience and keep on pointing out the vast years they have can be very discouraging. It's also hugely patronizing as they are implying that they were born with the knowledge and they weren't once young and also learning their way around the business.

15.2.7 Lack of Training, Coaching, and Mentorship

The STEM careers typically swing between not having enough time to train (or mentor/coach) as there is too much work, or not having enough money to train due to a lack of work. This can include technical and managerial training. My best advice for all of the above is to take young people to meetings with you. It costs you (practically) nothing and the experience they get cannot be obtained in a classroom. It's also about the old adage: rather train them and they leave, than not train them and they stay.

15.2.8 Limited Promotion

Managers often like to make assumptions about females regarding their capacity to handle family and a career, or the pressure of being in management. If you consider that these females are often good enough to handle the pressures of financial targets, project budgets, deadlines, and work quality, why would they not be able to handle promotion? Why do managers also assume in this day and age that females take on all the responsibilities for child rearing? Using the same logic, all young men that plan to be fully committed to supporting their partners in a parent partnership (as is the norm nowadays) should also be questioned of such plans before having children and not be promoted in case they can't cope.

15.2.9 Lack of Career Focus and Passion

Often in tough times, we take whatever job we can get. Or, if it's your first job, you are just so grateful to get the job experience. You may move departments to try out other fields, but you are still not finding your passion. Years go by and although you are not happy, you stay because you have to pay the mortgage or school fees, or help dependents. It takes a brave soul to admit the unhappiness, but an even braver soul to make a change.

15.3 INFLUENCES ON WOMEN IN ENGINEERING

15.3.1 The Strongest Positive Influences and Support

The strongest positive influences for women in STEM are their parents and spouses; however this can alternatively be grouped under the term "support system" for more unconventional families/communities.

> This support system is of utmost importance, as one of the hardest issues facing women is coping with sick children and coping with dropping and collecting children from school/activities. This is in addition to having to assist with homework and projects, cook, clean, and ensure all required shopping is done such as groceries, kid's clothes, and other ad hoc requirements.

Having to put up your hand as a woman in a meeting/workshop to announce you have to leave/leave early due to issues surrounding children is one of the biggest causes of stress and frustration for women as it often results in a feeling that your commitment is questioned. In contrast, should a man leave for the same reason he is viewed more favorably as he goes "above and beyond" for the family.

Having a strong support system to assist during these situations is invaluable and will go a long way in reducing the stress created, also creating a backup system should women need to travel for work.

Luckily, I can say that I have fantastic male colleagues who not only assist wholeheartedly, but some cook exclusively and they all view parenting as a partnership. On a personal note, I have to add that my husband, Gerrie, could not be more supportive as a partner in parenting, and I feel I am the luckiest mom in the world to have such a great, supportive family and the best friends who are always there when I am in a bind.

15.3.2 The Strongest Negative Influences

The lack of support at work and home, perceived discrimination or impediments to opportunities, and policies and practices specific to women (pregnancy) are the strongest negative influences to women in STEM.

Women in STEM already find work circumstances challenging due to the aspects mentioned above, but combined with the guilt of leaving their children in the care of others (often strangers) for the whole day, it is very easy to see why women in STEM leave the workforce or struggle with managerial positions if they are continuously treated aggressively by others. If you continue not treating women in STEM as equals, why should they stay and make you money? Research shows the benefits in cashflow and revenue you enjoy because of gender parity, so just support back!

15.4 WHAT DO WOMEN IN STEM NEED?

Women in STEM need

- Flexible working conditions
- Cheaper or company-supported childcare (women want to work and feel fulfilled after having children, so don't just make assumptions around maternity leave)—oh and have conversations with grown-ups!
- Pay parity

- Policies relating to the work–life balance
- Generally, more company and management support
- Proper graduate training programs
- Job stability and continuity

15.5 PRACTICAL TIPS AND TRICKS AS A WORKING WOMAN AND MOM TO HELP WITH A BALANCED LIFE

- There is *no such thing as a perfect woman*, so ask for *help*. You need a supportive partner and, if at all possible, a maid/helper/mom/friend. A good family support system will also help, especially when kids get sick. And they will! A lot!
- *Outsource all nonessential tasks* that fit within your budget. You don't need to bake that birthday cake from scratch. You are allowed to buy pre-cut pumpkin. Have your groceries delivered. Make food that you just pop in the oven or roasted in the slow cooker during the day. Make life as easy as possible for yourself!
- Focus on your *productivity at work*: Do you really need to take work home or can you improve the speed at which you work, spending less time on trivial issues?
- *Multitasking* is a skill that can be developed and improved, and as a female it's built into your DNA; otherwise most kids wouldn't make it to adulthood.
- *Put family first*, put family first, put family first! Don't skip on family events, take photos! These moments are precious, and life goes by so fast. With proper planning you can put family first and still make your deadlines.
- *Hug and kiss your loved ones.* It's been scientifically proven to reduce your stress.
- *When you get home, "BE" home.* Stop thinking about work when you are at home, work is not thinking about you. You need to shut off when you get home in order to be more productive at work, and your family deserves it, too. You will miss out on what your kids or family has to tell you. Play with the kids on the carpet, at their level where they are assured you are involved. Only when the kids are in bed, carry on working (but only if required—because you are going to improve your planning and productivity, my penguin!).

- *Spending special time with friends and family* is essential to give you perspective, laughter, and relaxation.
- *Lots of lists, for everything.*
- *Start work at 7 am* if your company has flextime; it helps you plan your day and allows more time with your family. It allows a proper break if you need to work again later in the evening. It also gets you home in time to cook, do homework, and bathe the kids.
- *Get your emails sorted* (possibly while having breakfast); only allow for periodic checking if you can. I know, I know ... but one can dream.
- *Have backup* if things run late at work (husband/friends/family/ helper).
- *Support others* (be someone's backup, shoulder, cheerleader).
- *Volunteer* (no really). This will be food for your soul when you need it during low points in your life.
- *Exercise* (but do what you enjoy). Try to get creative by incorporating exercise with family time, such as jumping on the trampoline with your kids. I know, I know, but one can dream ... again ... after two minutes I too almost die.
- *Sleep!* As much as you can, limiting digital light. If you really struggle to sleep, speak to your doctor about taking a light sleeping pill. There is no shame in wanting good, deep sleep.
- *Limit social media.* It can be very toxic. It can also be very embarrassing. So, if you must, use it for the greater good and spreading positivity.
- Have a *"work uniform."* It reduces a serious amount of decision making on a daily basis, leaving scope for better decision making at work. Make it easy to travel with; for example, make sure it doesn't crease and matches your one pair of black shoes and black jacket. And Barack Obama does it ... so there!
- Try to *eat as healthy as possible*, ensuring you and your family get enough vitamins through food. Minimize processed food that will leave you feeling tired. Don't be shy about taking supplements if you need them to support your health and stress levels.
- When nothing else works and it all gets to be too much, *have a glass of wine.*

The secret to having it all is knowing that you already do!

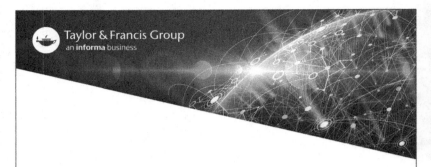

Printed in the United States
by Baker & Taylor Publisher Services

Printed in the United States
by Baker & Taylor Publisher Services